Life Beyond Earth

Luigi Vacca

Life Beyond Earth

The Fermi Paradox—Or Why We Are Still Waiting to Meet Aliens

 Springer

Luigi Vacca
L'Aquila, Catania, Italy

ISBN 978-3-031-81694-9   ISBN 978-3-031-81695-6   (eBook)
https://doi.org/10.1007/978-3-031-81695-6

This Springer imprint is published by the registered company Springer Nature Switzerland AG
The registered company address is: Gewerbestrasse 11, 6330 Cham, Switzerland

If disposing of this product, please recycle the paper.

Dedicated to my father, Angelo

Preface

This book is about life outside Earth. It is written for the general public. While the book contains lots of technical information, its main scope is to spark the imagination of our readers thanks to its intriguing hypotheses. I took the liberty of mentioning scenarios that border on science fiction. I wrote it so it would be also fun to read. The book was written to cover as many topics as possible, giving the reader a flavor of some of the problems in astrobiology. This keeps the reader from getting bored with excessive technical points. It was also written to cover aspects of life on Earth, including our history, in a comparative analysis while limiting the book's size to less than 200 pages. It is easy to imagine that alien life forms share many features that distinguish living things on Earth.

The book is divided into two different parts: the former is dedicated to Enrico Fermi, the universe, and the Drake equation, while the latter part is entirely devoted to some of the major hypotheses that can explain the Fermi paradox. The Drake equation is a heuristic equation that estimates the number of communicating civilizations in our galaxy. The equation consists of seven parameters: some of these parameters are of astrophysical, astronomical nature while others are of chemical, biological nature. The book discusses different cases for the number of advanced civilizations by varying the values of a couple of parameters. According to a Los Alamos report, the Fermi paradox was born when Fermi was having lunch with his colleagues in 1950 and discussing advanced extraterrestrial beings' existence. Just like Fermi, I find it strange that despite our significant efforts, we have yet to see any evidence of extraterrestrial life. The problem of the existence of alien life is one of the many scientific and philosophical questions for which we seek definitive answers.

Some people may wonder why a novice, albeit with an old plasma physics degree, would write a book about this important subject. Indeed, my current line of work is artificial intelligence. I wrote it because I am curious about our universe and the mystery of life. I am a little bit of a philosopher. I started studying the subject a while ago, puzzled by my intuition telling me there must be intelligent life elsewhere. As I age, I ask myself the existential questions that everybody asks. I am confident that we will one day answer Fermi's question. Over the years, I decided to put my ideas into writing this book while realizing the hardness of this task because I was a non-specialist.

Understanding my limitations, I sought the feedback of many specialists, but I have received little or none. In any case, I never got discouraged. Writing forced me to study new subjects to widen my views while searching for answers. Ultimately, I was pleased by the positive reaction when I presented it to Springer editors. I worked hard to answer their questions and meet their requirements.

Even though I am not a specialist in the field and the book covers many different fields, I have made a good effort to make sure that the technical parts are correct. Please let me know if you find something incorrect, and I will include it in the next edition.

Please let me thank an anonymous referee for their constructive feedback and suggestions. Furthermore, I would like to thank Angela Lahee at Springer for advising on this process and cosmologist Jennifer Wagner for her comments on my chapter on the universe.

L'Aquila, Italy   Luigi Vacca, Ph.D.
October 2024

Contents

1

What Is the Fermi Paradox?

1.1 The Scientist

Who was Enrico Fermi? Enrico Fermi was an outstanding Italian physicist who made numerous discoveries and inventions in nuclear energy and quantum theory. He was born on September 29, 1901, in Rome, Italy. He was the third child of Alberto Fermi and Ida de Gattis. His father was a division head in the Ministry of Railways (Ferrovie). His mother was an elementary school teacher.

From a young age, Enrico Fermi demonstrated exceptional aptitude in physics. His early interest in the subject was a coping mechanism for the loss of his brother Giulio during surgery. Fermi's command of mathematics allowed him to delve into advanced physics, such as the treatise on mechanics by Simeon-Denis Poisson, a French physicist, and the general physics textbook by Orest Chwolson, a Russian physicist. Enrico Fermi's exceptional preparation and dedication to physics led to a significant victory in an academic competition. This victory earned him admission to the prestigious Scuola Normale in Pisa, founded by Napoleon in 1810 to support gifted students based on merit. During his studies at the Normale, Fermi learned the theory of relativity and quantum mechanics as an autodidact. He was well respected at the Normale. Even professors asked him about the latest discoveries in physics.

At the remarkably young age of 21, Fermi earned his doctorate from the Scuola Normale in 1922, defending a thesis on the phenomenon of diffraction produced by X-rays. The following year, he secured a fellowship from the Ministry of Public Education and journeyed to Göttingen, Germany, to study under Professor Max Born, a luminary in quantum mechanics. Fermi also met with Werner Heisenberg and Wolfgang Pauli, two of the fathers of quantum mechanics. However, feeling somewhat dissatisfied with this arrangement, he

L. Vacca, *Life Beyond Earth*,
https://doi.org/10.1007/978-3-031-81695-6_1

relocated to Leiden in the Netherlands in 1924 to collaborate with Paul Ehren-fest, an eminent physicist renowned for his work in statistical mechanics and quantum theory. This association proved immensely inspiring for Fermi, reaf-firming his immense potential. This collaboration with Ehrenfest is no coinci-dence since Ehrenfest sent a letter to Fermi in 1923 to ask him questions about Fermi's latest work on ergodic theory. At Leiden, Fermi was honored to meet Albert Einstein, a Swiss-American physicist who is considered by many to be the greatest physicist ever [1]. In 1925, he returned to Italy and was invited by the mayor of Florence to teach physics and mechanics at the University of Florence for two years.

In 1926, Fermi discovered the so-called "Fermi-Dirac statistics," which gov-ern gas particles that act according to the Pauli exclusion principle. Nowadays, these particles are called "fermions" in Fermi's honor. Examples of fermions are the electron, the proton, and the neutron. It is also to be mentioned that Paul Dirac, an eminent British physicist, discovered the Fermi-Dirac statistics independently, hence the name of the statistics. In 1927, Fermi secured the prestigious full Professor of Theoretical Physics position at the University of Rome. In Rome, he assembled a group of physicists who would research the intricacies of radioactivity and the atom's nucleus with him. Due to their young age, this group is forever known as "I ragazzi di Via Panisperna" (the Boys of Via Panisperna). The group members were Franco Rasetti, Emilio Segré, Edoardo Amaldi, Bruno Pontecorvo, and Ettore Majorana. The Institute of Physics, where their experiments took place, was located on Via Panisperna [2].

In 1933, Fermi published his theory on the beta decay [3]. Armed with Wolfgang Pauli's theory of neutrinos, Fermi advanced the theory that during the transformation of a neutron into a proton, an electron and a neutrino would be generated and emitted. Fermi's work opened the door to the treatment of the weak force, one of the four fundamental forces of nature.

The group dedicated themselves to investigating the effects of neutron bombardment on materials, a research endeavor that would prove pivotal in paving the way for the groundbreaking discovery of nuclear fission by Ger-man chemists Otto Hahn and Fritz Strassmann, along with Austrian-Swedish physicist Lise Meitner [4] in 1938. In 1934, Fermi realized slower neutrons were more effective in inducing radiation on the bombarded elements. He had the intuition to use paraffin, a lower-atomic substance, rather than lead to slow neutrons. However, Fermi and his group missed that bombarding heavier elements, such as uranium, with neutrons yielded the formation of elements with lower atomic numbers, such as barium. Nevertheless, Fermi's experiments on induced radiation received much attention from the physics world.

In 1938, the Royal Swedish Academy of Sciences awarded Fermi the Nobel Prize for his work on the bombardment of elements by slow neutrons. Following his trip to Stockholm to receive the Prize, Fermi emigrated directly to the United States, driven by the urgent need to safeguard his Jewish wife, Laura Capon, from the horrors of fascist persecution. Laura Capon was the daughter of Admiral Augusto Capon, who tragically perished in Auschwitz in 1943.

Fermi arrived in New York by boat in January 1939. He was offered a professorship at Columbia University, where he conducted fission experiments.

Subsequently, Fermi went to the University of Chicago, where he spearheaded the construction of the inaugural fission reactor, colloquially termed the "graphite pile," employing graphite as a medium for neutron moderation. In the graphite pile, a nuclear fission chain reaction was taking place. A neutron collides with the nucleus of an atom, which causes that nucleus to split into two new, lighter nuclei. The split nucleus releases additional neutrons, which split other nuclei, and so on. The graphite pile was the first example of a nuclear power reactor. The first nuclear reactor went critical on December 2, 1942, a moment that can be considered the beginning of the nuclear age.

Fermi joined the cohort of scientists in 1944, developing the first atomic bomb, the "Manhattan Project," which culminated with the successful explosion of the first nuclear bomb on July 16, 1945. After the end of World War II, Fermi worked as a professor at the University of Chicago while spending some time in Los Alamos, New Mexico, as his expertise was required. He also served as a consultant to the Atomic Energy Commission. He continued to do physics research focusing on particle physics and cosmic radiation while forming a new generation of physicists. Sadly, Fermi's life was cut short by cancer in 1954 at the age of 53.

1.2 Where Is Everybody?

A Los Alamos National Laboratory report by Eric M. Jones provides evidence of Enrico Fermi exclaiming the famous phrase: "Where is everybody?". The reference pertains to the existence of intelligent life on other planets. According to the testimonies of scientist friends Edward Teller, Herbert York, and Emil Konopinski, Fermi exclaimed the famous words during a summer lunch in 1950 at Los Alamos in New Mexico.

According to Konopinski, when he joined his friends for lunch, he found they were discussing flying saucers. The discussion then moved on to the possibility that flying saucers could exceed the speed of light to travel in interstellar space. Indeed, it is widely recognized that according to the special theory of

Fig. 1.1 Fermi at the blackboard. By Smithsonian Institution—Flickr: Enrico Fermi (1901–1954), Public Domain, https://commons.wikimedia.org/w/index.php?curid=12686013

relativity, no material object can match or surpass the speed of light. There-fore, discussions about flying saucers from outer space were viewed purely as speculative, given the limitations imposed by this fundamental principle of physics.

At the end of the lunch, Fermi exclaimed the famous phrase, "But where is everybody?". Fermi discussed estimates of the number of planets such as Earth in the universe, the likelihood of intelligent life on such planets, and the consequent technological development that could lead alien beings to visit Earth in the future. From these estimates, Fermi concluded that aliens should have visited Earth many times.

Fermi's question continues to spark tremendous interest and remains a focal point of extensive research. The Fermi paradox emerges from the vastness of the universe, teeming with countless stars and planets, juxtaposed with the absence of evidence regarding the existence of extraterrestrial life forms possessing intelligence and technologies akin to or surpassing our own.

This book explores various potential resolutions to the Fermi paradox. According to the author, the paradox stands as one of the most significant inquiries from both scientific and philosophical perspectives, persisting as an enigma that still puzzles researchers.

2

The Immensity of the Universe

2.1 What Is the Universe Made Of?

The universe is filled with energy that can take on different forms. For instance, matter is energy. There may be two distinct types of matter in the universe: ordinary matter and dark matter. Ordinary matter can form structures like gas clouds, moons, planets, stars, and galaxies. Dark matter does not interact with radiation but exerts a significant gravitational force in the universe. Radiation is also energy. Light is electromagnetic radiation emitted by hot bodies. Light is characterized by its frequency or wavelength. For instance, the cosmic background radiation is microwave radiation. Solar radiation covers a range of frequencies, from radio waves to gamma rays, including visible light that illuminates Earth. The composition of frequencies in a wave is called "spectrum."

Understanding the scale of the universe necessitates grasping the concept of the speed of light. Light, an electromagnetic wave, traverses the vacuum of space at a constant speed of roughly 300 thousand kilometers per second. To give an idea of the speed of light, a ray of light emitted by the Earth reaches the Moon in just over a second, while the light from our Sun takes about 8 min to reach the Earth. Based on these examples, it would certainly appear that the speed of light is breakneck, if not instantaneous. This perception arises because, in human terms, 300 thousand kilometers represents an immense distance, far exceeding any distances we encounter on Earth. However, when considering the scale of the universe and cosmic time, we easily conclude that the speed of light is quite slow, as we shall soon find out.

Now, let us imagine light traveling through space for an entire year. Light covers an astonishing distance of about 9.46 trillion kilometers in one year! This mind-boggling distance, known as a "light-year," is a commonly used

© The Author(s), under exclusive license to Springer Nature Switzerland AG 2025
L. Vacca, *Life Beyond Earth*,
https://doi.org/10.1007/978-3-031-81695-6_2

unit of distance in astronomy. Another popular distance astronomers employ is the "parsec," equivalent to 3.26 light-years.

There are two ways to get a sense of the size of the universe: (A) counting the number of galaxies, stars, and planets in it, and (B) estimating its observable boundaries by collecting the faint light that reaches us from the farthest objects. Let us begin with the number of galaxies.

2.2   The Number of Galaxies in the Universe

A galaxy is an immense gathering of gas, dust, planets, and billions of stars. Estimates regarding the number of galaxies in the observable universe vary. It was initially thought the universe counted as many as 2 trillion galaxies based on NASA's Hubble Space Telescope surveys [6]. Tod R. Lauer, Marc Postman et al. have analyzed measurements of the cosmic optical background made by the camera of the space probe New Horizons [7]. The space probe has traveled to the outer solar system to avoid the masking effect of solar light. The analysis of the measurements has proved that optical light comes only from known galaxies. This result suggests that the actual number of galaxies in our universe may be smaller than the estimate from the Hubble telescope by an order of magnitude. However, some galaxies are too faint or too distant from us to be observed. The James Webb Telescope can observe galaxies by operating in the infrared spectrum. Therefore, future estimates of the number of galaxies may vary. Even with a lower estimate, given the vast expanse occupied by galaxies and the interstellar voids between them, these numbers depict a reality that stretches beyond the bounds of our imagination.

Within each galaxy, an abundant array of stars can be found. Smaller galaxies may host fewer than a million stars, whereas larger ones can boast hundreds of billions, if not trillions, of stars. Our galaxy, the Milky Way, harbors an estimated 100 to 400 billion stars [8] and spans a distance of approximately 100,000 light-years from end to end. It's evident that no exhaustive tally of stars has been conducted; these figures are derived from extrapolations based on studies of celestial regions.

2.3   The Number of Stars in the Universe

Now, by multiplying the estimated number of galaxies by the average number of stars in a galaxy, we get a sense of the total number of stars in the universe. This is probably anything between 10^{22} and 10^{24} [9].

Let us take a brief detour to discuss cosmological numbers. In cosmology, exceedingly large numbers are frequently represented using exponential notation.

For instance, 10^{22} can be read as the number 1 followed by 22 zeros: 10000000000000000000000.

Given the immense scale of these numbers, it would be easier to contextualize them by comparing them to familiar objects or concepts. For instance, if we were to hypothetically gather a million Earths and meticulously count every grain of sand covering their surfaces, the resulting count would be comparable to the top estimate for the total number of stars within the observable universe.

2.4   The Number of Planets in the Universe

What is a planet? The typical definition is a non-luminous body orbiting a star. It is spherical and not an asteroid. Furthermore, there should not be any significant debris on its path.

However, this definition is fraught with problems because it has been discovered that an incredible number of planets wander in space without being gravitationally tied to a star, the so-called nomad or rogue planets. According to Louis Strigari and his collaborators, nomad planets far outnumber ordinary planets by orders of magnitude [10]. However, due to their distance from any star, they do not reflect enough light and, therefore, are extremely challenging to detect and observe. Some nomad planets may have originated as planets orbiting stars but were ejected from their solar systems. Gravitational interactions caused the ejection during close encounters with other stars or planets or to chaotic instabilities in their parent stellar systems' orbits.

Traditionally, our solar system was recognized to contain nine planets: Mercury, Venus, Earth, Mars, Jupiter, Saturn, Uranus, Neptune, and Pluto. However, in 2006, an international consortium of astronomers, the International Astronomical Union, reclassified Pluto as a dwarf planet. Mercury, Venus, and Mars are among the rocky planets composed primarily of silicate rocks and metals. In other solar systems, there are super-Earths, rocky planets with solid surfaces at least twice Earth's size but smaller than Neptune. Furthermore, there are Neptune-like planets in our system known as "ice giants," which include Uranus and Neptune. Being distant from the Sun, Uranus and Neptune are characterized by extreme coldness. They are rich in frozen water, methane, and ammonia on a solid core, earning them their designation. Finally, Jupiter and Saturn complete the roster; these are massive planets predominantly composed of gaseous elements such as hydrogen and helium and are often called "gas giants."

What about planets outside our system that are tied to stars?

As of 15 July 2024, scientists and researchers confirmed the discovery of 5,690 planets orbiting other stars, known as 'exoplanets' [11]. This number is expected to rise significantly as the search for exoplanets expands. One method employed to detect exoplanets is the transit method. This approach involves observing the reduction in the brightness of a star when a planet passes between the star and the Earth, known as a transit. Due to the relatively small size of planets compared to their parent stars, the decrease in brightness is typically subtle, often amounting to just a few percentage points of the total luminosity emitted by the star. Furthermore, the method requires that the planet transits between the star and the Earth to be recorded. Consequently, this method tends to identify planets relatively close to their parent star and of substantial size, as they can obstruct a more significant percentage of the star's luminosity.

TESS (Transiting Exoplanet Survey Satellite) is an MIT-NASA mission to find candidate exoplanets using the transit method, which has found thousands since its launch in 2018.

Another method to discover exoplanets is the Doppler shift method. As commonly experienced, the frequency and wavelength of sound from a moving ambulance change as it approaches or recedes from us.

This phenomenon applied to light from stars and galaxies is termed "redshift" when the object moves away from us and "blueshift" when it moves closer. In astronomy, a similar principle is applied to identify exoplanets by observing periodic variations in a star's light spectrum. These variations are caused by the gravitational tug of an orbiting planet, inducing a slight wobble in the star's orbit.

How many planets exist in the universe? Determining the total number of planets in the observable universe remains challenging, given our limited understanding of exoplanets. Astronomers believe there is at least one planet per star in the Milky Way [12]. This number stems from the search for exoplanets, which has shown that there is more than one planet per star. There could be more planets than the search has shown since planets with smaller sizes and more considerable distances from their star are challenging to discover.

2.5   Moons: Small But Promising!

Moons, celestial companions gravitationally tethered to planets, offer captivating prospects as potential cradles for life. While Earth has just one moon and Mars claims two, the gas giants Jupiter, Saturn, Uranus, and Neptune boast an astonishing collective tally of over 200 moons, as reported by the National Aeronautics and Space Administration (NASA) [14]. Jupiter has 95 moons, according to the latest 2024 NASA count, while Saturn has 146 moons, more than any other planet. Moon counts are subject to constant revision and updates.

Several of these moons have subsurface oceans and thin atmospheres. These moons originated from gas disks encircling their parent planets. Exomoons, the moons beyond our solar system, remain elusive due to their low mass, rendering detection challenging. Consequently, our knowledge of exomoons remains scant, though their allure lies in the prospect of liquid water–a potential harbinger of life.

Moons, typically bodies gravitationally bound to planets, should be carefully considered in the search for extraterrestrial life. Researchers at Columbia University led by astronomer David Kipping are actively looking for exomoons [13]. Given what we know about our solar system, we should expect more exomoons than exoplanets in the universe. However, this statement remains speculative because exomoons are hard to detect.

The presence of water on the large moons of our solar system suggests that the probability of primitive life on exomoons, if any, may not be negligible. The moons of giant planets may be of the right size and gravitational field to host life, whereas a giant planet may not.

2.6 The Age and Size of the Universe

2.6.1 The Properties of the Universe

The cosmological principle states that the universe is homogeneous and isotropic. A homogeneous universe is equivalent to saying that when observing the universe from a different galaxy, galaxies and the clusters in the sky look uniformly distributed at large scales and on average, as seen from here. Isotropic means that by changing the orientation of our telescope, the universe looks about the same. The prevailing idea is that the cosmological principle is valid at large scales of hundreds of million light-years [16]. It is essential to add that the cosmological principle is a subject of current debate and research among cosmologists [17–19]. Furthermore, let us define what constitutes the "observable universe." It is defined as the spherical space region around the Earth from which we can receive light. Every observer in the universe occupies the center of their particular observable universe. The maximum distance light has traveled to reach us since the universe began is what defines the observable universe's boundary. A quantity named "particle horizon." From now on, we will always refer to the observable universe unless specified otherwise.

2.6.2   Edwin Hubble's Discovery

The starlight that illuminates our sky comes from stars in their distant past, thousands, millions, and even billions of years ago. The frequency or wavelength of light can be used to discover the elements that emit it. Wavelengths are, in a certain sense, the fingerprint of elements that we find in nature.

The American Vesto Slipher used a spectrograph to study the emission and absorption of light from galaxies (then called "nebulae"). From the analysis of light from galaxies, he observed that the spectra were quite similar to those of objects on Earth, but they were shifted. In a series of articles from 1913 to 1917, Slipher documented shifts in the light spectra of galaxies [20–23]. Slipher's research paved the way for the discovery that the universe is expanding.

In 1929, the renowned American astronomer Edwin Hubble provided experimental evidence indicating that the universe was expanding. Utilizing the most advanced telescope of his era, a 100-inch reflector located on Mount Wilson in California, he made a groundbreaking discovery: galaxies situated farther from us exhibit faster recession velocities [24]. To arrive at this result, Edwin Hubble selected a series of distant galaxies and measured each one's radial velocity and distance from Earth. The radial velocities were measured as Vesto Slipher did. Once he obtained such data, he plotted the velocity of galaxies on a graph as a function of their distance. He observed that a straight line interpolated the data points on the graph. This is equivalent to saying that there is a linear relationship between the radial distance of a galaxy and its velocity; such a relationship is called Hubble's law. The current rate of this expansion, velocity divided by distance, is known as the Hubble constant or parameter.

How did Hubble measure the velocity of the galaxies using the Doppler shift principle? We have already seen the use of the Doppler shift method to search for exoplanets. Hubble recorded the spectral emission of the galaxies and observed a redshift in the lines of the light spectrum of the galaxies. The presence of a redshift means that a galaxy is moving away from us. The amount of redshift determines the radial velocity of the galaxy. The measurement of the distance of a galaxy is more challenging. In the case of distant galaxies, astronomers use "candles," which are stars whose luminosity is either known or assumed. Direct distance measurements are only possible for close objects at distances of a few thousand light-years. Essentially, the distance of a candle can be computed from its known luminosity and apparent brightness and the application of the inverse square law. Another type of candle is a type 1a supernova; such objects are typically assumed to have the same luminosity. However, such objects are typically rare.

Hubble focused on a particular type of star: a Cepheid variable. These stars pulsate radially, changing their brightness. One can derive their luminosity from measuring the period of such stars. From their luminosity and their apparent brightness, Hubble computed their distances. Using the same technique, Hubble determined that spiral nebulae, Slipher's nebulae, are, in reality, galaxies outside the Milky Way.

Hubble underestimated the distance of the galaxies he studied by a factor of seven due to inaccuracies in the calibration of the Cepheids. Precise measurements of the Hubble constant were made using the Hubble telescope, named after the great American astronomer. To complicate matters in cosmology, measurements of the Hubble constant differ depending on the method used to measure it, the so-called "Hubble tension." More precisely, there is an apparent discrepancy between the higher Hubble constant value measured using supernovae 1a and Cepheids and the smaller one obtained from the Cosmic Microwave Background radiation from the early universe. The Hubble tension is one of the most challenging problems in cosmology.

2.6.3   The Universe is Expanding

The galaxies' increasing radial velocity with distance indicates that the universe's space is expanding. An observer in any galaxy will see another galaxy receding and vice versa while the space between the two galaxies expands. Let us take the two-dimensional example of dots on a balloon. It can be shown that Hubble's law is consistent with the cosmological principle. In particular, the expansion may occur at all scales but is not so apparent at small scales. Galaxies and what is within remain mostly unchanged because their gravitational attractive force predominates over the expanding force. For instance, the distance between the Earth and the Sun is not appreciably modified by the expansion process because of the gravitation force the two bodies exert on each other.

An interesting observation is that Hubble's law predicts that the radial velocity of galaxies exceeds the speed of light when galaxies reach a given distance for an observer. The minimum distance where galaxies equal or exceed the speed of light is the Hubble distance. A careful reader should know that the special theory of relativity is not violated in this case; galaxies are not traveling locally faster than a ray of light passing by; rather, it is the space that separates that is expanding.

General relativity explains Hubble's law and its results, such as expanding space and superluminal expansion. What do we know about gravity in general? A quick description of the special and general theories of relativity is in order.

2.6.4   The Theory of Relativity in a Nutshell

The special theory of relativity was published in 1905 by Albert Einstein. This theory describes the effect that the finite speed of light has on space and time [25]. Einstein's theory is better understood using the formulation of spacetime by the mathematician Hermann Minkowski, Einstein's former professor at the Swiss Federal Institute of Technology in Zurich [26]. Special relativity postulates that the laws of physics are invariant in inertial frames and that the speed of light in a vacuum is the same for all observers. An inertial frame is a frame that is either stationary or moving at constant velocity. Under these conditions, Einstein's special relativity explains the effects of time dilation and length contraction between inertial frames. It is assumed that the observer is in an inertial frame. Time dilation occurs when an observer sees a slower clock moving away from her. Similarly, an observer observes an object's length contraction when the object is moving away in a different inertial frame. Such laws were initially derived before 1904 by Dutch physicist Hendrik Lorenz. Hence, they carry Lorenz's transformations because Lorenz was the first to formulate both laws. Einstein's special relativity includes such laws into a general framework. One significant result of his theory is that massive objects can never achieve the speed of light because doing so would require infinite energy. One of the major results of this theory is the equivalence of mass and energy given by Einstein's equation that relates mass to energy $E = mc^2$, where E is the energy, m is the mass, and c is the speed of light. The factor c^2 explains how nuclear reactions can produce enormous energy. However, the special theory of relativity does not explain gravity (Figs. 2.1 and 2.2).

In 1915, Albert Einstein published his theory of gravity or general theory of relativity [27]. His theory posits that gravity is caused by space and time's bending (curvature) due to energy-momentum density and flux through the spacetime continuum. Einstein's field equations express such a relationship. Once a metric of the spacetime continuum is obtained from a solution of Einstein's equations, matter and light follow the shortest paths in the four-dimensional space described by the metric. Such paths are called geodesics. For instance, a geodesic path on a sphere is a great circle. In general, the geodesic path is determined by the spacetime geometry. A falling apple follows the path determined by the geodesic of the spacetime curvature determined by the presence of the Earth's mass. In the words of American physicist John

Fig. 2.1 Albert Einstein during a lecture in Vienna in 1921. By Ferdinand Schmutzer/Adam Cuerden—http://www.bhm.ch/de/news_04a.cfm, (2006) archived copy (image), Public Domain, https://commons.wikimedia.org/w/index.php?curid=34239518

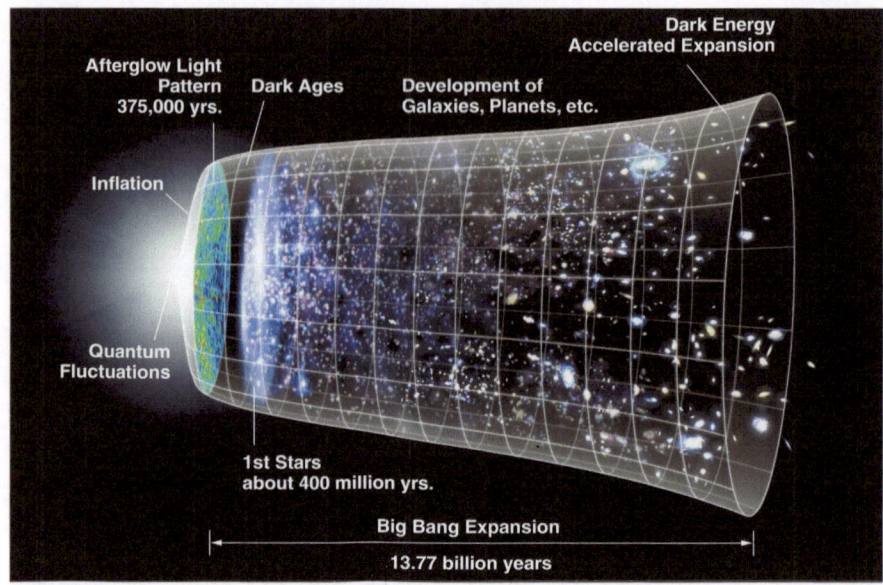

Fig. 2.2 Timeline of the universe. A representation of the evolution of the universe over 13.77 billion years. By NASA/WMAP Science Team—Original version: NASA; modified by Cherkash. Public Domain, https://commons.wikimedia.org/w/index.php?curid=11885244

Wheeler: "Space-time tells matter how to move; matter tells space-time how to curve."

Einstein believed in a static universe, so he added a constant to his field equations: the cosmological constant. The cosmological constant added the repulsion force that prevented the universe from collapsing due to the force of gravity [28]. However, Einstein's model was unstable. A tiny change in matter density in his model leads to a continuing expansion of the universe or its collapse. Fourteen years after the publication of the theory of general relativity, the Hubble discovery proved Einstein's conception of the universe wrong. Einstein himself called the cosmological constant "his greatest blunder."

The theoretical basis for Hubble's discovery existed before the discovery was made. In 1922, a Soviet mathematician and cosmologist, Alexander Friedmann, formulated equations describing an expanding universe as a solution to Einstein's field equations [29]. In particular, he inserted a scale factor representing the temporal evolution of the universe's expansion. The dynamics of the scale factor are governed by the mass-energy density of the universe and its geometry.

2.6.5   Dark Energy and Dark Matter

In 1998, a relatively recent observation of supernovae showed that the universe expansion is accelerating [30–32]. The acceleration of the expansion was discovered by Saul Perlmutter, Brian Schmidt, and Adam Riess, for which they were awarded the Nobel Prize in 2011. Daniel Eisenstein and his collaborators independently corroborated the acceleration in a study based on the data from the baryonic acoustic oscillations. These are fluctuations of the density of baryonic matter, matter made of quarks [33]. Based on their best models and empirical observations, some cosmologists think that a mysterious form of energy called dark energy accelerates the universe's expansion. No one knows the origin and nature of such energy. Quantum mechanics tells us that any vacuum has energy, and therefore, vacuum energy could be a candidate for explaining the acceleration of space expansion. However, quantum calculations of vacuum energy density don't yet explain dark energy density; indeed, the vacuum energy density is higher than the dark energy density by an enormous factor of 10^{120}. The consensus states that this form of energy is pervasive in the universe at a cosmic scale, constituting 68% of all the mass-energy density in the universe.

Another form of matter in the universe whose nature is unknown has been hypothesized: dark matter. This type of matter does not interact with ordinary matter or radiation except through its gravitational field. Dark matter was discovered when Swiss-American astronomer Fritz Zwicky observed insufficient matter to keep a spinning cluster of galaxies from flying apart [34]. A result that was confirmed by Vera Rubin and Kent Ford in the 1970s on individual galaxies [35]. Dark matter is believed to be about 26% of the total mass-energy density in the universe. The remaining five percent is due to the presence of ordinary matter and radiation we experience daily. Therefore, most of our universe is still pretty much a mystery.

2.6.6   The Big Bang Theory

A Belgian Catholic priest, Georges Lemaitre, starting from Einstein's equations, derived a theory of an expanding universe that supported Hubble's law. Einstein pointed out that Lemaitre's theory followed Friedmann's work [39]. Starting from his result, Lemaitre hypothesized that points in space that are now far apart were much closer in the past until they converged back to almost coincide. Lemaitre's idea is the foundation of the "Big Bang" theory.

It is now widely acknowledged that the universe originated from a densely compacted primordial state approximately 13.8 billion years ago. Such a primordial state comprised subatomic particles such as quarks and gluons. Quarks are the building blocks of protons and neutrons, while gluons are responsible for the strong nuclear force.

As the universe started cooling, hydrogen and helium atoms formed. Over billions of years, gravity formed gas, dust, galaxies, stars, planets, and all the objects that populate our universe. All probes hint at 13.8 billion years as the most probable age of the universe after assuming that the universe's expansion has accelerated. Furthermore, after including cosmic expansion, the universe's age is consistent with the ages of the oldest stars. Such stars must certainly be younger than the universe. One of the most fundamental discoveries in the field of cosmology followed Hubble's findings more than 30 years later: the cosmic microwave background radiation.

2.6.7   The Cosmic Background Radiation

In 1964, American physicist Arno A. Penzias and radio astronomer Robert W. Wilson, while working at Bell Labs in New Jersey and studying radio emissions from our galaxy, made a remarkable discovery for which they received the Nobel Prize in Physics in 1978 [36]. They observed that regardless of the direction they turned or how they adjusted their equipment, their detector consistently measured a constant level of radiation coming from all parts of the microwave spectrum. Theorists supporting the Big Bang theory deduced that this radiation was the residue of the hot early universe about 400,000 years after the Big Bang. This radiation is universally known as the Cosmic Microwave Background Radiation (CMBR) [37]. The Planck space mission showed us that the CMBR is incredibly uniform all over the sky with tiny temperature variations, just as Penzias and Wilson found out. In addition, the CMBR has a blackbody spectrum at a temperature of 2.7 K [38], much colder than the temperature at which it originated. Traveling across an expanding universe for more than 13 billion years, the CMBR wavelength became more and more stretched, resulting in a cooler temperature.

2.6.8   A Flat Universe

Studies of the CMBR performed by the space mission WMAP tell us that the universe is spatially flat within a percent [40]. Such results have been confirmed by the data from the Planck surveyor mission, which aims to measure the anisotropies of the CMBR [41]. The CMBR tells us that the early universe must have been very flat, a flatness that requires the universe's energy density to be perfectly equal to a given value according to the general theory of relativity, i.e., the critical value for energy density. The cosmological inflation theory can also explain this coincidence. A universe with a density higher than the critical density will collapse, while a universe with a lower density than the critical density will expand forever.

2.6.9 Cosmic Inflation

One of the most accepted theories of the universe in its very early stage is the theory of cosmic inflation. It was introduced by Alexei Starobinsky, Alan Guth, Andrei Linde, and others to explain a series of problems, among which is the uniformity and smoothness of the temperatures of the CMBR in the context of the Big Bang [42–44]. It was hard to see why parts of the universe could have reached thermodynamic equilibrium so effectively in the early stages.

To explain this problem, cosmological inflation posits that in the very early universe, an infinitesimally small section of the universe, much smaller than a proton, expanded exponentially by a factor of 10^{26}, driven by metastable vacuum energy density. The inflationary phase is believed to have lasted 10^{-32} seconds. The section expanded to the size of an apple faster than the speed of light; following this period of rapid expansion, inflation started fading away. The exponential expansion smoothed out the region, explaining the CMBR's properties and our universe. After the end of the inflationary period, the universe has continued to expand for billions of years, and in the last 4 billion years, dark energy has even accelerated this expansion.

2.6.10 The Size of the Universe

Light emitted during the phase of the early universe traveled almost 13.8 billion years to reach us, which may suggest that the radius of the universe is 13.8 billion light-years. However, this is a wrong conclusion because space has expanded since then. As we have seen, this process makes mega clusters of galaxies fly away from each other. Hence, the distance from the farthest object we can see since the universe's birth is not just given by the universe's age multiplied by the speed of light but by how far such object has moved away from us due to the universe's expansion. This distance is the size of our observable universe, about three times the space covered by light since the universe's beginning, or 46.5 billion light-years [45].

2.7 Summary

The universe is unimaginably vast, with countless moons and planets, many of which may harbor life. Furthermore, there is an unobservable universe of which we know nothing and could be infinite. With this in mind, one might wonder: are we alone in the universe? And can we say something about the likelihood of primitive life in our galaxy, given our level of scientific knowledge? What about the possibility of intelligent life? This intriguing question introduces us to the renowned Drake equation, which we will delve into in the next chapter.

3

Drake's Equation

3.1 Who Was Frank Drake?

Frank Donald Drake was a famous American astrophysicist and astrobiologist [46]. Born in Chicago on May 28, 1930, he was fascinated by the idea of the existence of extraterrestrial civilizations from a young age. Subsequently, he enrolled at Cornell University, receiving a BA in Engineering Physics. Drake served as an electronics officer on a naval warship. His passion was astronomy. Hence, he enrolled in a Master's and a PhD program in astronomy at Harvard, which he earned in 1955. He began his career as a radio astronomer in 1958 at the National Radio Astronomy Observatory in Green Bank, West Virginia, and worked there until 1963. At the observatory, he initially concentrated on studying planets and pulsars. His projects included studies of Jupiter's radiation belts, Venus's atmosphere, and mapping the Milky Way's center. He contributed to extending the capabilities of the Arecibo Observatory to capture radio signals.

In 1959, at Green Bank, Drake started "Project Ozma," searching for radio transmissions from advanced extraterrestrial civilizations. Project Ozma's observations began on April 8, 1960. In September 1959, before Project Ozma, physicists Giuseppe Cocconi and Philip Morrison published a paper in Nature titled "Searching for Interstellar Communications." The authors proposed searching for radio signals from intelligent extraterrestrials in the article. For Drake, this publication was a further incentive to continue his project.

In the end, no aliens were detected, but a graduate student at Cornell named Carl Sagan, who later became a renowned American astrobiologist, contacted Drake. This led to lifelong cooperation between the two astronomers in search of extraterrestrial civilizations.

© The Author(s), under exclusive license to Springer Nature Switzerland AG 2025
L. Vacca, *Life Beyond Earth*,
https://doi.org/10.1007/978-3-031-81695-6_3

In 1961, at a scientific gathering at Green Bank discussing the potential existence of advanced extraterrestrial civilizations within our galaxy, Frank Drake penned his now-famous Drake equation [47]. He intended to spark a dialogue among fellow attendees. Little did he realize, perhaps, that his equation would captivate the attention of scientists worldwide, becoming the focus of intense study in the coming years by astrobiologists and fascination among non-experts.

He was a professor at Cornell University from 1964 until 1984 and the director of the Arecibo Observatory. In 1972 and 1973, Frank Drake, Carl Sagan, and Linda Sagan drew the design of the Pioneer Plaques for the missions Pioneer 10 and 11. This was the first example of a physical message sent into space to initiate contact with aliens. In 1974, Drake wrote the Arecibo message in binary code, broadcasted toward the Great Globular Cluster in Hercules, 25,000 light years away. Also, in 1974, Drake was elected as a fellow of the American Academy of Arts and Sciences. In 1977, he worked with Carl Sagan to send the Golden Record, a collection of photos and audio, on the Voyager mission. Drake was a member of the National Academy of Science and worked on the Phoenix Project for SETI, a comprehensive search for extraterrestrial intelligence, from 1995 to 2004. In 1984, he became a Professor of Astronomy and Astrophysics at the University of California, Santa Cruz. Drake passed away in 2022 at 91 years of age in his house in California (Fig. 3.1).

3.2   Probability and Independent Events

Drake started with a probabilistic argument to arrive at his famous equation.

Let us now embark on a mathematical parenthesis to provide readers with a flair for quantitative analysis with some foundational concepts in probability. While this section isn't imperative for grasping the Drake equation, it provides valuable insights for those interested in its mathematical aspects and theoretical aspects [48–53]. The study of probability seeks to assign numerical values to the likelihood of random, non-deterministic events. Examples of random events can be found everywhere in nature; it is enough to think of the random Brownian motion of suspended particles in a liquid. However, it was gambling and the financial markets, two fields that have initially spurred and continue to spur the study and development of probability theory.

Consider a scenario where an individual seeks to estimate the probability of winning the lottery and a poker game. One straightforward method involves calculating the instances in which individuals achieved this dual victory, then dividing it by the number of times people participated in both the lottery

Fig. 3.1 Dr. Drake revisited the variables of the Drake Equation several decades after its inception. By Raphael Perrino—Flickr: Dr. Frank Drake, CC BY 2.0, https://commons.wikimedia.org/w/index.php?curid=23566349

and poker across all potential combinations. This method aligns with the frequentist approach to probability.

An alternative method for estimating probabilities is the Bayesian approach, named after its originator, the English reverend Thomas Bayes. In 1763, Bayes published a seminal work on probability theory titled "Essay Towards Solving a Problem in the Doctrine of Chances." Within this treatise, Bayes introduced his renowned formula for conditional probability, which we include here due to its practical importance. Formally, when considering two events A and B, the probability of both events occurring together is contingent upon the conditional probability of event A given B, as expressed by Bayes' formula:

$$P(\text{A and B occur together}) = P(\text{A occurs given event B}) * P(\text{B occurs})$$
(3.1)

This formula can be extended to multiple events using recursion. In the case of events A, B, and C is given by:

$$P(\text{A, B and C occur together}) = P(A) * P(\text{B given A}) * P(\text{C given A and B}) \qquad (3.2)$$

and so on.

Finally, always in the case of statistical independence, the probability of multiple events co-occurring is straightforward: it is the product of all individual probabilities of each event.

For example, for independent events A, B, and C, the probability that they occur together is given by the product of the individual probabilities as follows: $P(A)P(B)P(C)$. Furthermore, in comprehending the Drake equation, it is important to recognize that we multiply the number of attempts by the probability of winning to determine the expected number of wins. For instance, if we play the lottery ten times, the expected number of wins is derived by multiplying ten by the probability of winning, assuming that the probability of winning each time is independent of the previous. With that, we conclude our digression on probability calculations. Now, let us introduce the Drake equation (Fig. 3.2).

3.3   The Drake Equation

The Drake equation is a heuristic equation that estimates the number of intelligent extraterrestrial civilizations in the Milky Way capable of communicating with us or any other intelligent civilization in our galaxy. It is written as a multiplication of parameters as follows:

$$N = R * f_p * n_e * f_l * f_i * f_c * L$$
(3.3)

Where:

- N is the number of extraterrestrial civilizations in the Milky Way with which some form of communication is possible.
- R is the average star formation rate in the Milky Way in a year.

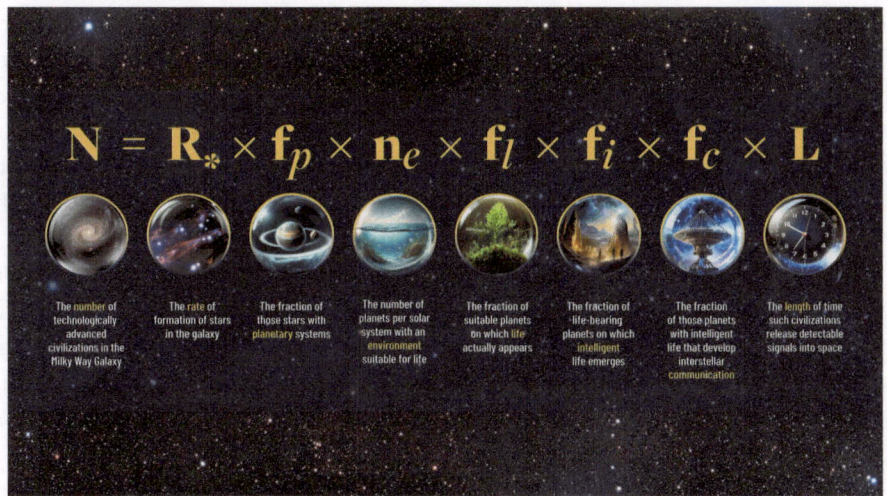

Fig. 3.2 The Drake Equation. By NOIRLab/AURA/NSF/P. Marenfeld—The Drake Equation, CC BY 4.0, https://commons.wikimedia.org/w/index.php?curid=147958110

- f_p is the fraction of those stars in the Milky Way that have planets orbiting them.
- n_e is the average number of planets that can potentially support life given a solar system with at least one planet.
- f_l is the fraction of life-supporting planets per solar system where life develops.
- f_i is the fraction of those life-supporting planets where evolution creates intelligent life.
- f_c is the fraction of extraterrestrial civilizations that produce technology capable of transmitting understandable information signals.
- L is the average yearly duration during which a technologically advanced civilization will transmit signals in the cosmos.

According to Bayes' formula, this equation can be analyzed in a heuristic fashion, giving us a feel for the equation [48]. First and foremost, $R * f_p$ is just an estimate of the yearly number of newly formed solar systems with planets in our galaxy. This number refers to the distant past in our galaxy, millions of years ago, for an intelligent civilization to have sent a message in their past. More generally, it can be interpreted as referring to any extraterrestrial civilization that wants to contact another. The other following terms are estimates that are conditional on their preceding parameter. For instance, n_e is the average number of planets that can support life conditional on being part of a solar system with at least one planet. Continuing, f_l is the fraction of planets that host life and are part of a solar system with at least one planet. The parameter f_i refers to the fraction of planets with intelligent life conditional on the fact that they had some life and that the planet could support life. For instance, we could have plenty of planets where life is purely microorganic and not intelligent. Finally, f_c deals with transmitting species given the chain of events that include the potentiality for life and the presence of simple or complex life on a planet in a solar system with at least one planet.

Alternatively, the Drake equation can be seen as a steady-state equation where civilizations are born, initiate contact, and are replaced by new civilizations after they perish. The expected yearly frequency of transmitting civilizations must be multiplied by the yearly duration of communication during which they launch signals into space, aligning with our initial utilization of the annual star formation rate.

3.3.1   The Problem of Intergalactic Communication

As the reader may have noticed, the original formulation of the Drake equation is confined only to our galaxy. Indeed, it is possible to formulate an equation for the observable universe. In this case, the number of planets with life will be astronomically more significant, which is undoubtedly a positive factor. One problem with this approach is that, presumably, the closest galaxy to us, the Canis Major Dwarf Galaxy, is an irregular dwarf galaxy 25,000 light years away from us [54]. It is now thought that Canis may be part of the Milky Way galaxy. More realistically, the Andromeda Galaxy is about 2.5 million years from Earth.

Any signal we might intercept today from Canis Major would have been transmitted long before the emergence of human civilization, predating the Neolithic era and the advent of the earliest agricultural societies. Hence, it is improbable that any communication sent by extraterrestrial beings inhabiting a galaxy beyond our own was intended specifically for us. In essence, attempting to send a message to a neighboring galaxy parallels tossing a message-bearing bottle into the vast expanse of the sea. It is also improbable that an extraterrestrial civilization would spend an immense amount of energy transmitting a signal over extragalactic distances, knowing that such a signal may never reach its intended target.

The final recipient of any intergalactic message, if anyone at all, may remain unknown to the sender or be a hostile party.

There is another possibility. Extraterrestrial beings may have already sent us a message, having foreseen the eventual emergence of intelligence on our planet from their studies of Earth as it was millions of years ago. When the message arrives millions of years later, no one may be left to read it.

For all the reasons above, communication among civilizations should be limited to our galaxy.

3.4   An Alternative Form of the Drake Equation

One possible criticism of the Drake equation is that it should be extended to include exomoons. If we did that, the number of advanced extraterrestrial civilizations could increase by a factor of ten or more. Another possible critique is that the Drake equation does not adequately represent the dynamics of life. Paths conducive to intelligence may be fraught with setbacks for life. It is hard to come up with reliable estimates for the biological parameters of the Drake equation.

Many authors provide alternative formulations and critiques of the Drake equation. For instance, Anders Sandberg, Eric Drexler, and Toby Ord critique the use of point estimates in the Drake equation [55]. According to the authors, point estimates in the Drake equation lead to a decent, if not even large, number of extraterrestrial civilizations. Such a result stems from the improbability of life on an exoplanet, which is balanced by the innumerable number of exoplanets and exomoons in our galaxy and our universe. Drawing a mathematical analogy is equivalent to multiplying zero by infinity and getting a finite number. Furthermore, they also observe that different estimates can yield vastly different numbers for the number of civilizations. Their recipe introduces uncertainty in each parameter using different historical estimates from different researchers. Once they have assembled a set of values for each parameter, the authors randomly extract a value for each parameter until every parameter in the Drake equation is sampled. The sampled values for the parameters are multiplied to obtain the number of civilizations. Repeating this exercise using random generators, a probability density function is obtained for the number of civilizations.

While not excluding the existence of other civilizations altogether, the authors conclude from their exercise that the *ex-ante* probability that we are alone is high. Therefore, according to the authors, one should not be surprised that we are not finding evidence of extraterrestrial life.

3.5   What Is Next?

The following chapters will provide rough estimates of all parameters except in two cases: the intelligence parameter and the communication duration. In the author's opinion, these parameters are the hardest to evaluate. It is enough to consider that there is only one technological species on the planet for millions and millions of different species to see why the task is complex. In the final chapter devoted to the Drake equation, our distinctive approach is to vary these parameters to find the relationship between the two unknown parameter values for a given number of civilizations.

Notwithstanding, a qualitative discussion of all the parameters will be provided. The Drake equation is typically used to estimate the number of communicating civilizations in the Milky Way. In all likelihood, point estimates for the Drake equation parameters are not only highly instructive but also a synthesis of the synergy between the physical universe and the biology of life. Distilling various sources of information into a single number for each parameter forces us to think critically and touch many fields, from mathematics to physics, astronomy, cosmology to chemistry, and biology.

4

Stellar Formation

4.1 What Is a Sun?

Our Sun is a star. Stars have been visible in the night sky well before the genesis of humanity and even before the appearance of the first life forms on Earth. But what is a star from a physics point of view?

A star can be described as a plasma furnace, generating new, heavier elements through the fusion of atomic nuclei, beginning with the simplest and most abundant element in the universe: hydrogen. Stars consist mainly of hydrogen and helium with small carbon, oxygen, nitrogen, and iron fractions. The Sun is made of 70% of hydrogen and 25% of helium and other elements. Plasma, the most prevalent form of matter in the universe, differs from solids, liquids, and gases by being ionized—a state where electrons have been separated from ions. When referring to plasmas within stars, one typically alludes to exceedingly high temperatures, often reaching tens of millions of degrees at the core of these celestial bodies, 15 million degrees Celsius in the case of our sun [56].

The nuclear fusion reactions occurring within a star's core are fueled by the immense gravitational pull resulting from the star's mass, generating intense inward pressures and temperatures within. Despite the repulsive forces between ions, known as the Coulomb force—a term honoring the pioneering work of French scientist Charles-Augustin de Coulomb in electromagnetism—the high velocities of ions within this plasma allow them to collide and merge, surmounting the Coulomb force. This collision and fusion process gives rise to heavier elements and liberates substantial energy per reaction, much more than a fission reaction, the splitting of an atom.

The merging of ions to create a higher atomic element is called "a fusion reaction."

L. Vacca, *Life Beyond Earth*,
https://doi.org/10.1007/978-3-031-81695-6_4

Stars are recognized in the sky by their brilliance. They are the essential components of a galaxy. Studying a star's lifecycle, from its inception to its demise, is crucial in unraveling how our universe will unfold.

4.2   Star Classification

Various stars exist, ranging in mass, element composition, luminosity, external temperatures, and spectra. Stars can be classified based on the kind of light they emit, temperature, size, and luminosity, among other criteria. There are different types of stellar classification. For instance, the Harvard stellar classification system, pioneered by American astronomers Annie Jump Cannon and Edward C. Pickering [57], classifies stars according to their surface temperatures. For instance, if we consider stars with higher mass and temperature, those much larger than the Sun fall into types O, B, and A, with type O being the hottest and most massive. These O-type stars appear blue to blue-white, boast temperatures significantly higher than our Sun, and are exceedingly rare. The luminosity of a star is proportional to its mass. O, B, and A-type stars collectively constitute less than one percent of all stars in our galaxy. Stars of types O and B typically have short lifespans, far shorter than the billions of years required for the emergence of intelligent life. Consequently, due to their rarity and short lifespans, they are of lesser relevance in our context. On the other hand, A-type stars typically have lifespans between 0.6 and 2 billion years, yet they remain relatively rare [58, 59].

Continuing down the classification hierarchy, we encounter yellow and white F stars, slightly larger and hotter than the Sun. F stars comprise approximately 3% of all stars within the Milky Way [60]. Following them are G-type stars, constituting around 7% of all stellar bodies [61]. Notably, our Sun belongs to this category, being a yellow G-type star. The classification sequence concludes with K and M-type stars, characterized by temperatures and masses more minor than that of the Sun and exhibiting an orange or faintly red hue. M stars predominate in our galaxy, constituting about 80% of all main-sequence stars in the Milky Way [62]. These lifespan statistics refer to a star's "main sequence" phase, where hydrogen fusion occurs, encompassing most of a star's lifecycle. For instance, a main sequence typically endures for about 10 billion years for an average star like our Sun, extending up to 100 billion years and even a trillion years for stars with smaller masses known as "red dwarfs" or typically M-stars, which, as previously mentioned, are the most prevalent type of star in the Milky Way. The careful reader may have noticed that these smaller stars can, in fact, last much longer than the current estimate of the universe's life! However, what is the natural course of life for stars like the Sun or those that are smaller?

4.3 The Lifecycle of Stars

Let's introduce some basic concepts about stellar formation. Stars originate from nebulae, where gravitational forces draw together clouds of interstellar hydrogen gas and cosmic dust. Typically, a star's evolutionary trajectory is determined by its mass, with higher-mass nebulae generally giving rise to higher-mass stars. The life cycle of stars with low and medium mass and temperature, such as our Sun, commences with the formation of a protostar. This protostar forms from a gas cloud, which, set into motion, begins to contract and condense. As the condensation progresses, hydrogen atoms within the protostar collide, elevating pressure and temperature until nuclear fusion initiates. During this fusion process, hydrogen atoms combine in the core to form heavier helium atoms, releasing substantial energy that fuels further fusion reactions. Consequently, a star operates as a vast, self-sustaining reactor. Throughout the main sequence phase, the inward gravitational force toward the star's core is counterbalanced by the outward pressure exerted by core nuclear fusion. This equilibrium persists as long as hydrogen fuel is available for fusion within the star's core. However, once hydrogen is depleted, the star transitions to fusing heavier elements, although this phase is relatively brief compared to the main sequence duration. Additionally, as higher elements are fused, their timescales diminish. Upon exhaustion of its hydrogen fuel, a star can no longer maintain equilibrium between gravitational forces and fusion reactions. Consequently, the star's core collapses inward, leading to the expulsion of its outer layers. This process results in the star's expansion into a "red giant," a fate anticipated for our Sun in approximately 5 billion years when it is projected to engulf even the Earth (Fig. 4.1).

During the red giant phase, a different type of fusion reaction occurs, wherein nuclei fuse to generate carbon and oxygen nuclei. Once the helium fuel is exhausted, the star transforms dramatically, expelling most of its mass as a cloud of material known as a planetary nebula. The residual core of the star, now diminished by the gas ejection, evolves into a compact object termed a "white dwarf." Over time, these white dwarfs gradually cool down and transition into an inert body.

This process represents the future for stars like our Sun in the Milky Way that do not have enough mass to become a neutron star or a black hole. Additionally,

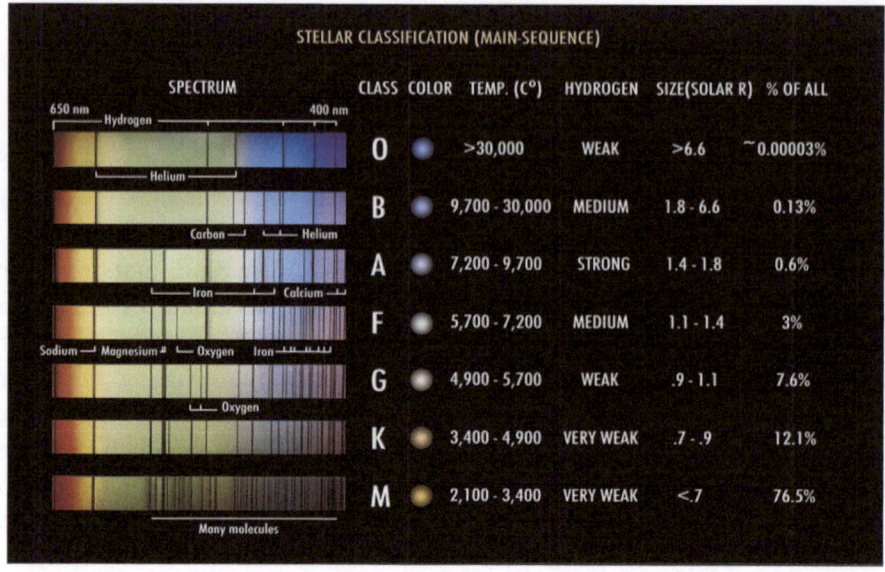

Fig. 4.1 Star types with their typical spectrum, color, temperature range, hydrogen lines abundance, size, and the fraction of all known stars. Pablo Carlos Budassi, CC BY-SA 4.0, https://creativecommons.org/licenses/by-sa/4.0, via Wikimedia Commons

it is crucial to discuss the evolution of stars with masses that are significantly bigger than our Sun. These stars burn brighter and have much shorter lifespans compared to lower-mass stars. They also progress through a sequence of fusion reactions to heavier elements. However, when they produce iron nuclei, fusion reactions no longer yield net positive energy, as dictated by nuclear physics. At this juncture, gravitational forces overpower fusion reactions, compressing the star's core to immense temperatures.

A spectacular explosion known as a supernova can occur when a massive star core is subject to enormous pressure and temperatures of billions of degrees Celsius. Most of the star's material is expelled into interstellar space during a supernova event. The expelled matter can enrich other systems with heavier elements that are part of our bodies. Following a supernova, if the remnant core's mass is sufficiently high, it can collapse into a "black hole," a singularity in spacetime, as per Einstein's general relativity. In 1915, German mathematician Karl Schwarzschild discovered the first solution to Einstein's field equations, a spacetime metric containing such a singularity. The singularity revealed the concept of a black hole—an incredibly dense object capable of trapping even light. Black holes were experimentally observed in the early 1970s [63, 64], but only recently, in 2019, direct observation of a black hole has been accomplished with the Event Horizon Telescope.

In cases of a lower mass core, massive stars form compact stars composed of densely packed neutrons, called "neutron stars." Notably, approximately half of all solar systems in the Milky Way contain two or more stars gravitationally bound together, unlike our solitary solar system. Systems wherein two stars orbit a common center of gravity are termed binary star systems. Like high-mass stars, binary stars can culminate in a supernova explosion, particularly when two heavy binary stars collide.

4.4   The First Parameter of the Drake Equation

As seen, stars undergo a life cycle reminiscent of the processes observed in living organisms. Like living beings, all stars are born and eventually meet their demise after millions and billions of years once their nuclear fuel is extinguished.

The first parameter in the Drake equation, denoted as R, refers to the birth rate of stars. Specifically, it represents the average rate of star formation occurring within the Milky Way galaxy billions of years ago. Several considerations must be made: between 13 and 6 billion years ago, there was a tremendous rate of stellar formation in the heart of our galaxy, followed by a quiet period. Hence, stellar formation is not uniform in space and time [65]. Furthermore, the galaxy's center is unlikely to have solar systems that host life because it is characterized by a supermassive black hole, Sagittarius $A*$, and supernova activity.

With this in mind, several studies offer estimates for R. A recent study shows that our galaxy produces between 0.5 and 6 solar masses every year [66]. Although this production is not the most active among the many galaxies in the universe, this number assumes significance in the Drake equation and is deemed scientifically reliable.

The pace of stellar formation in our galaxy is diminishing due to the depletion of cold gas reserves.

We can conservatively estimate the magnitude of the first parameter as of the order of unity, as Drake and his colleagues previously estimated. Such an estimate is most likely conservative after considering the variability in time and space of the rate of space formation.

$$\boxed{R \approx 1} \tag{4.1}$$

4.4.1 The Question of Life in R

Nevertheless, not every star within the Milky Way is compatible with the sustenance of life. What fraction of these newly formed stars during the main sequence is likely to have conditions favorable for the emergence of life?

This question remains open-ended, with only a narrow scope for response. The earliest life forms emerged on Earth approximately 3.7–3.8 billion years ago, as the discovery of sedimentary formations generated by microbial life in Greenland suggests [67].

The inception of the technological era of humanity occurred approximately two centuries ago—a notably brief span when juxtaposed with the existence of life on Earth. Consequently, it is plausible that the emergence of an extraterrestrial civilization boasting a technological prowess on par with or surpassing our own would necessitate billions of years commencing from the ground up.

As we have seen before, the smaller the stars, the longer they last during their main sequence. Therefore, it can be inferred that stars with masses similar or smaller to that of our Sun can be conducive to supporting life. That is the vast majority of main-sequence stars in the Milky Way.

4.5 Types of Stars and Life

Giant stars (type O and type B, for example) last much less than the Sun and emit an ultraviolet spectrum that is very harmful to DNA. Hence, they should not be conducive to systems with planets that host life.

Small stars, such as those of K and M types, often emit significant solar flares, posing potential hazards to life. Consequently, in theory, they are generally considered less ideal for hosting life-supporting solar systems. However, recent studies show that the planets in an M-type solar system can host life due to the protective effect of ice sheets for tidally-locked planets [68]. Conversely, medium stars in the main-sequence phase, like our Sun, are deemed more compatible with life, primarily due to their longer lifespans. Stars of type F or even A could be included in the list of stars compatible with life, as their extended lifespans span billions of years. Thus, if one were to estimate the parameter R based solely on the most favorable star types for life, such as G-type stars, it might amount to significantly less than unity. However, it is pertinent to note that other crucial parameters influence a star's compatibility with life. One such factor is a sizable habitable zone within the star's solar system. Planets situated too close to a star would be excessively hot for life, while those too distant would be too cold.

4.6   Carbon-Based Life in Extreme Environments

On the other hand, these arguments presuppose that any extraterrestrial civilization would share biological similarities with ours and that the prevailing conditions are not excessively extreme. However, special life forms have been discovered on Earth, thriving in extreme environments and temperatures. These life forms are called extremophiles and can be of type bacteria, archaea, or eukaryotes.

They can be classified into different categories according to their adaption [69]. For instance, thermophiles and hyperthermophiles survive in high-temperature hydrothermal vents. An example of a thermophile is the *Bacillus stearothermophilus*, a bacterium.

Psychrophiles are extremophiles that grow best at low temperatures [70]. Many of these psychrophiles are bacteria. However, some examples of eukaryotes are in this category, such as lichens and snow algae. Snow algae can be found in polar regions and thrive in the presence of snow, hence the name.

Other types of extremophiles are acidophiles and alkaliphiles [71]. Some acidophiles can live in extremely acidic environments (pH of 0). Alkaliphiles are organisms, typically bacteria, adapted to surviving in alkaline environments (pH of 11). Barophiles (piezophiles) are organisms that grow best under pressure. These organisms are typically bacteria found 10 km deep inside the ocean (Mariana Trench). Finally, we have halophiles, organisms that thrive in environments with high salt concentrations.

Many of these extremophiles are polyextremophiles; they survive in environments with several extreme parameters.

The existence of extremophiles suggests that life may thrive in environments that were once considered unthinkable.

This quality of surviving in extreme environments also applies to animals. Tardigrades, known as water bears due to their bear-like appearance, exemplify remarkable resilience [72]. These minuscule aquatic creatures, typically less than a millimeter in size, boast eight legs and can endure extreme conditions. They survive temperatures well below freezing and exceeding boiling, inhabit the depths of the Mariana Trench, and can tolerate extended periods without food. In 2007, tardigrades were exposed to the vacuum of space and intense solar radiation for 12 days. Upon their return to Earth, scientists found that most tardigrades had survived the ordeal, thanks to their ability to enter a state of suspended animation known as a "tun state." This illustration demonstrates that life forms reliant on DNA can endure environments far less hospitable than those found on Earth. Likewise, extraterrestrial life may also thrive in solar systems with stars different from our Sun.

4.7 Summary

With these considerations in mind and given our limited current knowledge, the R parameter should be somewhere of the order of magnitude of 1.

$$\boxed{R_{life} \approx 1} \qquad (4.2)$$

5

Solar Systems with Planets

5.1   Solar Systems with One or More Planets

We now move to the second parameter of the Drake equation f_p. This parameter represents the probability that a sun has at least one planet. As we have seen in preceding chapters, the universe and our Milky Way galaxy have a significantly greater number of planets than stars, conceivably by an order of magnitude.

With an excellent degree of certainty, it is generally accepted that the second parameter f_p is, most likely, close to unity [73]. This means that almost every main-sequence star in the Milky Way should have at least one planet, on average. For instance, the Kepler-90 star system has eight planets, like the Solar System. Indeed, it is a star system that closely resembles our system. The planets were discovered by NASA's Kepler space telescope thanks to the transit method. One of the eight planets, Kepler-90i, was discovered with the help of artificial intelligence applied to the transit method. The system has an F-type star, about 2800 light-years from Earth. The discovery of a solar system with many planets supports the idea that planets are much more numerous than stars in our galaxy. We would also do well to point out that this estimation excludes nomad planets, which are subject to being gravitationally captured by a nearby star. There is no chance that nomad planets may host life. On a very speculative side, radioactive decay can be a source of energy inside these planets that allows the existence of microorganisms. Incidentally, nomad planets and asteroids may play an essential role in the survival of advanced extraterrestrial lifeforms. Some of them may be gravitationally captured by stars and lead to originate life. In a highly speculative scenario, they could be significant for technologically advanced civilizations seeking refuge and access

© The Author(s), under exclusive license to Springer Nature Switzerland AG 2025   **37**
L. Vacca, *Life Beyond Earth*,
https://doi.org/10.1007/978-3-031-81695-6_5

Fig. 5.1 Kepler-90 system. By NASA/Ames Research Center/Wendy Stenzel— https://photojournal.jpl.nasa.gov/jpeg/PIA22193.jpg, Public Domain, https://commons. wikimedia.org/w/index.php?curid=64818984

to free kinetic energy. Such civilizations might consider temporarily relocating to nomad planets to evade predictable cosmic cataclysms, such as a supernova occurring dangerously close to their home planets. While our focus has been on planets in the context of Drake's second parameter, it's essential to recognize the potential importance of moons and other celestial wanderers in the search for extraterrestrial life. This acknowledgment piques curiosity and encourages further exploration. Let's now turn our attention to moons in our solar system (Fig. 5.1).

5.2 Life on Moons

As it is well known, lunar expeditions have yet to find any organisms on the Moon. This underscores the complexity and depth of our search for extraterrestrial life. It is widely accepted that the Moon is devoid of life. However, within our solar system, some moons have the potential to harbor life. Moons that can, in principle, support life are Europa, Enceladus, Titan, Calisto, Ganymede, Io, Triton, Dione, and Charon. Notably, the JUICE space mission, initiated by the European Space Agency in 2023, aims to investigate three moons orbiting Jupiter: Europa, Callisto, and Ganymede [74]. With its potential to uncover new insights, this mission holds great promise for understanding the potential for life beyond Earth. It's an exciting time for space exploration.

Additionally, another mission is being planned in the years to come to explore Titan and Enceladus (Explorer of Enceladus and Titan), the two moons of Saturn, where the conditions for life are believed to exist. Even Triton, the moon of Neptune, is of interest and may be subject to exploration in the future.

The optimism surrounding these moons stems from their icy surfaces, which likely conceal liquid water oceans beneath. The interaction between the icy crust and the moon's hot interior facilitates liquid water formation. This internal heat, generated by geological, chemical processes, and radioactive decay, is fueled by tidal forces from the massive planets to which these moons belong: Jupiter, Saturn, and Neptune. These giant planet's substantial gravitational forces on their moons contribute to their geological activity.

Water, as evidenced on Earth, is fundamental to the emergence of life. Simple forms of life could be swimming in the oceans of the moons of the giant planets. Furthermore, observations have detected water vapor in Europa, Enceladus, and Titan atmospheres, further bolstering the case for potential habitability [75–77].

Recent discoveries, drawing parallels to Earth, challenge previous notions about the limits of life in icy environments. Scientists from Aarhus University, led by Professor Alexandre Anesio, have found microbial communities thriving within glaciers [78]. These glaciers are teeming with microorganisms such as algae, bacteria, fungi, and viruses, adapting to the extreme cold conditions of polar environments. For instance, microbes within the Greenland ice sheet have adapted to polar temperatures, emitting significant amounts of volatile organic material in the process [79]. Due to their limited size, "exomoons," moons in other solar systems, have not yet been directly observed. Discoveries of life in icy environments bode well for life on exomoons. If this holds, the likelihood of discovering life forms within the Milky Way may surpass our current estimates based solely on the number of planets.

Moreover, besides liquid water, moons must maintain stable orbits around their respective planets for extended periods. This necessity arises from the notion that the natural evolution of life likely requires billions of years to give rise to highly complex organisms. Another critical factor is the mass of a moon, which contributes to the potential existence of an atmosphere. Essentially, every celestial body requires a minimum mass to retain an atmosphere through gravitational attraction. Thus, moons with masses comparable to Earth's are more promising for hosting life than their less massive counterparts. A protective magnetic field may surround moons with an iron core. Io, Europa, Ganymede, and Calysto have magnetic fields. Ganymede's magnetic field is generated internally by a dynamo effect in its iron core, while the Jovian magnetosphere induces other moons' fields.

However, although in an elementary form, moons and planets are not the sole objects theoretically capable of supporting some life components.

5.3   Meteorites and Life Building Blocks

A meteorite is an object that survives going through Earth's atmosphere. Meteorites are celestial bodies that partially survive atmospheric entry and collide with Earth, primarily due to their larger mass. Meteorites can be chunks of asteroids, comets, moons, or planets. They can be composed of rock, iron, and other elements. Millions of solid objects hit the Earth; fortunately, most are very small. For instance, the so-called "meteoroids" are typically small rocks that wander in space.

Space objects that reach kilometer-scale sizes are termed "asteroids" and can pose a significant threat to planetary life. Among the notable asteroid impacts on Earth is the Chicxulub asteroid, which struck Earth approximately 66 million years ago, leaving behind a crater 100 km wide. This cataclysmic event generated towering megatsunamis hundreds of meters high and ejected immense amounts of ash and debris into the atmosphere, triggering widespread fires and disrupting global ecosystems. The resulting dust particles shrouded the planet's atmosphere, leading to a prolonged period of reduced sunlight and subsequent flora depletion, ultimately contributing to the demise of the dinosaurs.

Despite their destructive potential, research suggests that meteorites could also serve as potential carriers of life components to planets or moons. Analysis of meteorite remnants has revealed that they may transport some or all of the essential DNA building blocks to Earth. For instance, we can cite the work of Yasuhiro Oba and his collaborators, who support this thesis [80]. These components may have contributed to or even accelerated the formation of life on Earth. The hypothesis that life or its primary components originated elsewhere in the universe and traveled to Earth transported by meteorites is called "panspermia." The idea of panspermia is not a new phenomenon. In the 5th century B.C., Greek philosopher Anaxagoras propounded the notion that life may have originated somewhere else and spread out throughout the entire universe [81].

An alternate theory that attributes the birth of life on Earth solely to a primordial inanimate soup of water, methane, and ammonia bombarded by strong ultraviolet radiation and lightning is known as "abiogenesis." Early theoretical proponents of the primordial soup theory were Soviet biochemist Alexander Oparin and British-Indian biologist J.B.S. Haldane, respectively, in the 1920s [82].

For a moment, let us entertain the possibility of the panspermia theory, suggesting that life's building blocks were deposited billions of years ago through interstellar meteoroids. Despite the vast distances between planets and moons in different solar systems, it is conceivable that numerous interstellar meteoroids could have transported life's components from one solar system to another over hundreds of millions of years. This process could be likened to an exogamous pollination on Earth, where wind aids in the dispersal of pollen. In this scenario, the spread of life may have propagated exponentially, traversing distances previously thought impossible across the galaxy and even throughout the universe. There are documented instances of meteoroids colliding with Earth, such as the case of the CNEOS 2014-01-08 meteor, which crashed in Papua New Guinea in 2014, as reported by CNN [83]. Such meteoroids were initially identified as interstellar, but it has been ascertained that they belong to our system. Identifying interstellar objects from those originating from our solar system presents a challenge since such occurrences appear to be rare. Given these factors, the panspermia theory is a plausible explanation alone or in conjunction with abiogenesis.

5.4   Summary and Estimates

In principle, planets, moons, and objects wandering in space can host life in its most varied forms, from elementary DNA components to technologically advanced alien civilizations. In the Drake equation, the parameter f_p refers only to planets, but as we have seen, this assumption is very conservative. A conservative estimate for the second parameter f_p of the Drake equation is that almost all star systems have one planet at the very least.

$$\boxed{f_p \approx 1} \tag{5.1}$$

6

Habitable Exoplanets

6.1 The n_e Parameter

Life, as we know it, could thrive on planets and moons with enough atmosphere, water, elements, and a protective magnetic field within a stable system. The potential for life in the universe extends beyond the confines of planetary bodies. Interstellar and solar system meteorites can contribute to spreading the blocks of life, albeit in simple forms. This broadens our perspective, suggesting that the probability of intelligent life in the universe is not limited to planets alone. It's a concept that invites us to explore and consider the vast possibilities of life beyond our current understanding.

Now, let's focus on the n_e parameter, a crucial factor in the Drake equation. This parameter represents the average number of potentially habitable planets in a solar system with at least one planet, underscoring its significance in understanding the likelihood of life beyond Earth.

What pivotal elements delineate a celestial body's suitability for habitation? The determinants are multifaceted and encompass many factors, some extending beyond conventional paradigms.

Here, we restrict ourselves to a few important factors necessary for life to occur [84].

1. The existence of a habitable zone.
2. The planet or moon composition and mass. The presence of a solid surface, an atmosphere, and water in liquid form.

L. Vacca, *Life Beyond Earth*,
https://doi.org/10.1007/978-3-031-81695-6_6

3. The type of a parent star.
4. The stability of its orbit.

Now, let's do a little deep dive into each one.

6.2   The Existence of a Habitable Zone

A habitable zone is a region of orbital space around a star where water on a planet's surface can be present in liquid form. This concept is not just important; it's crucial, as it defines the potential for life in our universe. The total amount of radiation that any planet or moon receives from its sun follows an inverse square law. A simple example can be made to illustrate the significance of this law. If we double the distance between a star and its planet, the total amount of sun radiation received by the planet is reduced by a factor of four. Hence, radiation wanes rapidly as the distance between a star and its planet increases. Intuitively, a planet or even a moon too close to its sun will have water present in gaseous form because of the very high amount of radiation energy that hits its surface. For instance, we have the case of Mercury, a scorching planet due to its proximity to the Sun [85], where water and other elements are in a gaseous state. Conversely, a planet too far from its sun will have water in a frozen state, as in the cases of Neptune and Uranus. Therefore, there must be an intermediate range of distances from a star that will permit liquid water. However, distance from a star is not the only factor. Indeed, a planet must have an atmosphere that exerts a minimum amount of pressure on the water to remain in the liquid phase. This is the case of our planet. Earth has an atmosphere that exerts significant pressure on surface water. The combinations of pressure and temperature that yield liquid water are found using a water phase diagram.

The determination of the inner edge and outer edge of the habitable zone in the solar system is a current subject of research. Here, we mention the results and rationale of a few research papers (Fig. 6.1).

6.3   The Earth Habitable Zone

An astronomical unit or AU is the distance between the Earth and the Sun, about 180 million kilometers. Illeana Gomez-Leal et al. studied a global climate model in the presence and absence of atmospheric ozone subject to increasing solar radiation [86]. The authors set the onset of the Moist Greenhouse

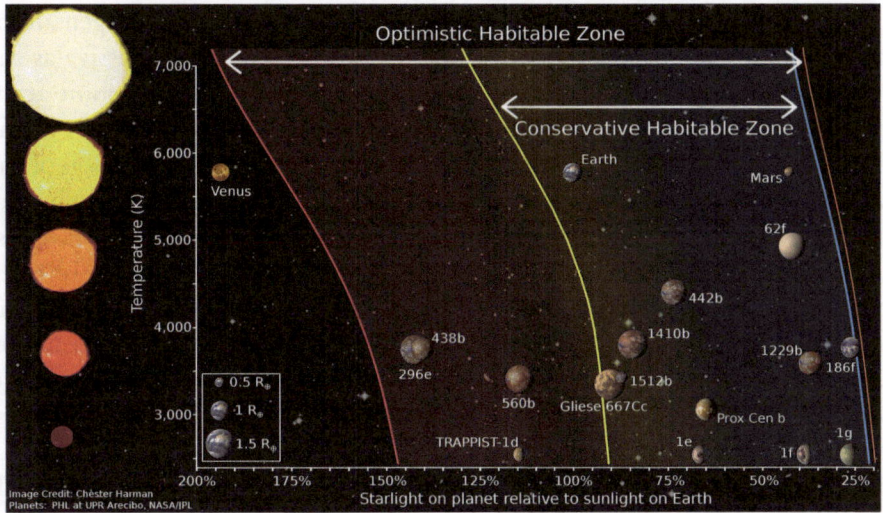

Fig. 6.1 This is the most widely used diagram for explaining the Habitable Zone. Shown is the temperature vs starlight received. Important exoplanets are placed on the diagram, plus Earth, Venus, and Mars. By Chester Harman.CC BY-SA 4.0, https://commons.wikimedia.org/w/index.php?curid=64107813

Threshold (MGT) when the Earth is 0.93 AUs from the Sun. MGT occurs when a significant amount of water vapor moves to the stratosphere, split into hydrogen and oxygen by solar radiation. Due to this separation, hydrogen leaves Earth, depleting it of water. Furthermore, the authors found the absence of ozone versus its presence increases the maximum solar radiation at which Earth stays habitable.

Jeremy Leconte et al. considers the problem of increasing solar irradiation over geologic times, consequent water evaporation, and its impact on the greenhouse effect [87]. Their use of a 3-D global model, instead of the 1-D models used before, shows that the threshold for a runaway greenhouse effect is higher than previously thought. One of the main findings of this study shows that the inner edge of the habitable zone for Earth is equal to 0.95 AUs and that Earth will not experience a runaway greenhouse effect due to increased solar activity for at least one billion years.

These results also apply to Earth-like planets and are useful in selecting potentially habitable exoplanets.

Ramses Ramirez and Lisa Kaltenegger show that the outer edge of the habitable zone of an Earth-like planet, precedently assumed at 1.7 AU in the solar system, can be extended to 2.4 AU due to the generation of atmospheric hydrogen due to volcanism. The assumption is that the release of hydrogen by volcanism will outpace the loss of hydrogen to space [88].

Pierrehumbert and Gaidos found that planets with a pure hydrogen atmosphere at 40 bars of pressure, when a planet is three times as massive as the Earth, can maintain the surface temperature above the freezing point at 10 AU for a star like our Sun [89]. The Earth, therefore, is situated closer to the inner boundary of the habitable zone in the Solar System. The location of the habitable zone also depends on the size of the star and, in part, on the physical properties of the exoplanet. The habitable zone is closer when we have less massive stars than the Sun. The zone is much further away in the case of more massive stars, such as O, B, and A-type stars. It should be added that stars generally do not have a constant luminosity even during their main sequence. On a narrow timescale, the Sun has an 11-year cycle during which the radiation emission goes from a minimum to a maximum [90]. Fortunately, at least in the Sun's case, current solar activity variations are not having a meaningful impact on our habitable zone's position. The Sun exhibits low levels of variability regarding its irradiance. The stability of the Sun's emission is an important factor in aiding the development of life on our planet. However, on timescales of billions of years, the Sun and similar stars will undergo a period of expansion and become brighter and brighter, becoming red giants and pushing the habitable zone further away.

A planetary orbit is also quite relevant as far as habitability is concerned. A planet with a highly eccentric orbit can go from extreme cold to extreme heat. Significant surface temperature gradients can be experienced by planets that are tidally locked to a star. The side of a tidally locked planet that faces a star is sterile due to extreme heat from the star, while the dark side is completely frozen.

Another crucial factor to consider, as exemplified by moons, is that the habitable zone can extend even further from a star when additional energy sources beyond solar radiation are considered. For instance, moons orbiting in elliptical paths can be heated by tidal gravitational forces resulting from interacting with a parent planet. Heat is produced within the moon as these forces cause expansion and contraction. Additionally, the energy emitted by radioactive elements and internal pressures may contribute to expanding the habitable zone.

6.4   Atmospheres in Exoplanets

We have already established the crucial role of an atmosphere in supporting life. Without it, water would either evaporate or freeze due to the absence of atmospheric pressure. Additionally, an atmosphere serves as a shield against harmful ultraviolet solar radiation. Consequently, a planet or moon must possess a minimum mass to retain an atmosphere gravitationally. For instance, our Moon lacks an atmosphere, while Mars has a thin one. Moreover, an atmosphere aids in storing heat from the star it orbits. Consequently, a planet with a larger mass may absorb too much heat, making excessively high surface temperatures unsuitable for life. Additionally, excessive gravitational force resulting from a larger mass could challenge life forms requiring mobility to survive. For instance, a planet with double Earth's radius could exhibit gravity forces two to three times greater than those on Earth, potentially favoring the presence of smaller species. Another critical factor in the presence of life is the presence of a solid surface. A rocky planet with a high content of silicate and carbon is more likely to have a solid surface that favors the presence of land animals and plants. As described by Archimedes' principle, ocean life experiences less downward force than land life due to water buoyancy, potentially accommodating larger species. It is not a coincidence that the largest animal on Earth is the blue whale. Therefore, the ideal planet or moon for organic-based life like ours should have a mass comparable to or slightly larger than Earth's. Rocky planets larger than Earth, with masses between twice and ten times Earth's mass, are termed "super-earths."

6.5   Exoplanetary and Exolunar Compositions

As seen in the previous section, the presence of an atmosphere is one of the fundamental requirements for life to appear and survive on a planet or moon.

The Earth's atmosphere comprises 78% nitrogen, 21% oxygen, and other elements such as carbon, hydrogen, neon, and argon. Therefore, at least theoretically, the presence of an atmosphere with a similar composition to Earth's could support life.

The presence of a robust magnetic field, generated primarily by a nucleus rich in iron, is another essential component for the emergence of life. This magnetic field shields Earth from the solar wind from the Sun's corona, which comprises ions, electrons, and alpha particles traveling at high speeds. These ionized particles pose a dual threat, damaging the Earth's atmosphere and living organisms by affecting the DNA in their cells.

Upon reaching Earth, these particles are deflected by the planet's magnetic field, protecting the entire ecosystem. Besides safeguarding against the solar wind, Earth's magnetic field shields us from cosmic radiation, which includes particles traveling at high speeds along with X-rays and gamma rays, known for their harmful effects on cells in large doses. Another significant heat source for Earth comes from the decay of radioactive elements, such as uranium, with high atomic numbers. This process heats the lithosphere and promotes tectonic activity.

Essentially, elements like water, hydrogen, oxygen, nitrogen, carbon, silicon, magnesium, iron, and many others are indispensable for the emergence and evolution of life. Rocky planets, abundant in rock and metal, typically contain all these essential elements, including radioactive ones.

6.6   The Type of Parent Star

As we have seen, the Sun is categorized as a G-type star. G-type stars, including our Sun, display surface temperatures ranging from 5,200 to 6,000 K during their main-sequence phases. Due to their radiation characteristics, these stars are likely the most conducive to supporting life. For instance, photosynthesis in plants relies solely on utilizing solar radiation within the visible part of the spectrum. The visible part of the spectrum goes from blue (higher energy) to red (lower energy). Plants appear green to the eye because they primarily reflect green light, with other colors being absorbed. Stars within the temperature range mentioned typically have lifespans of about 10 billion years, which is crucial for the evolution of life and requires such extended durations. These stars also boast well-defined habitable zones.

Conversely, these stars constitute the minority in our galaxy. K- and M-type stars are far more abundant, with M-type stars comprising the majority of all stars [91]. These stars are not deemed suitable for life due to their significant solar flare activity. However, M-type stars possess several advantageous characteristics: they have masses that are a fraction of our Sun's and remain on their main sequences for significantly extended periods than our Sun. Because of their lower mass, their luminosity is lower than the Sun's, resulting in less concentrated radiation in the spectrum's ultraviolet part. Due to their low luminosity, their habitable zone is narrower than our solar system and much closer than one astronomical unit. Studies indicate that planets with stable atmospheres and liquid water may still orbit M-type stars [92]. However, to be within the habitable zone, a planet would need to orbit relatively close to an M-type star and could potentially become tidally locked. As seen above, tidally locked planets consistently present the same side to the star, which could be vulnerable to intense solar flares, posing challenges for sustaining life. A similar consideration applies to K-type stars. The potential habitability of solar systems containing K- and M-type stars remains a significant focus of astrobiology research.

6.7 Orbital Stability

The gravitational stability of a planet for periods comparable to the duration of its star's main sequence is essential for the development of life. The stability of a solar system arises from the interplay of gravitational forces between all the planets and their suns.

6.7.1 Chaos Theory

To grasp the concept of dynamical stability, it's essential to say a few things about chaos theory. This theory, pioneered in 1972 by Edward Lorenz, an American mathematician and meteorologist, originated from his attempts to forecast weather patterns over extended periods. Lorenz aimed to solve hydrodynamic equations using a computer [93].

He observed that a slight variation in rounding a decimal number in the computer printout resulted in significantly different weather predictions than the previous scenario. In a nutshell, chaos theory asserts that even deterministic systems, which should theoretically be entirely predictable regardless of the timeframe, eventually exhibit quasi-random behavior. In other words, planetary bodies can become chaotic. This typically happens in nonlinear dynamics, where minor imperceptible effects have giant impacts over extremely long periods. It is entirely possible that a minute adjustment to an object's trajectory, such as a deviation of just a meter, could lead to vastly different trajectories over hundreds of millions of years or even more extended periods.

6.8 The Stability of the Solar System

The reassuring news is that according to numerical calculations, our solar system is projected to remain stable for at least another 100 million years. Mathematicians Angel Zhivkov and Ivaylo Tounchev from Sofia University in Bulgaria conducted a simulation demonstrating that the planetary configuration within our solar system should maintain stability over this extended timeframe [94]. However, some solar system arrangements exhibit greater instability than ours. For instance, binary star systems may introduce zones of gravitational instability for planets depending on the relationship between the two stars and the planets.

Another significant factor influencing stability is the eccentricity of a planet's orbit, which measures the deviation of the orbit from a perfect circle. Planets like Earth have minimal eccentricity, as their orbits are nearly circular. Planets with highly eccentric orbits experience dramatic temperature variations between their closest and farthest points from their star, potentially posing challenges to the development of life.

6.9   An Estimate for the Parameter n_e

Currently, estimating the average number of planets capable of supporting life remains largely an academic exercise, lacking significant empirical data. More than 5,000 planets outside our solar system have been discovered and confirmed. Ongoing studies focus on identifying which of the confirmed exoplanets are habitable [95]. Therefore, if we assume that solar systems have an average of 10 planets, the probability of finding a solar system with a habitable planet is $\frac{1}{10}$ on average. This estimate does not include the fact that Venus and Mars may have been habitable at some point. Habitable planets will be predominantly rocky, with masses exceeding Earth's but not substantial enough to classify them as gas giants. This prompts us to make two straightforward observations: (1) the list of exoplanets is statistically insignificant compared to the total number of planets in our galaxy, and (2) a good deal of these exoplanets are much more massive than the Earth.

One challenge in our quest to find exoplanets capable of sustaining life is that search techniques, like the transit method, tend to favor the detection of massive exoplanets orbiting close to their parent stars. Some estimates for n_e are much more optimistic and maintain that at least one planet per solar system can host life. Turning our focus to our solar system, we see that both Mars and Venus were very likely habitable in the past. Some theories suggest that Venus might currently host some form of microbial life within its clouds, supported by the detection of phosphine on the planet. The presence of phosphine and its source is still the subject of research [96]. Additionally, it is well-established that billions of years ago, Mars possessed an atmosphere. Liquid water flowed across its surface [97]. To conclude, we can hypothesize that the prevalence of habitable exoplanets in the Milky Way is proportional to the frequency of discovered habitable exoplanets multiplied by the average number of planets within a solar system, resulting in a number approximately equal to one-tenth. A number close to the frequency of habitable planets in our solar system.

$$\boxed{n_e \approx 0.1} \tag{6.1}$$

In a nutshell, on average, a solar system with ten planets has one habitable planet. More research is needed better to understand the composition and variety of exosolar systems.

We opine that this estimate is very conservative, suggesting that there could potentially be a significantly greater number of habitable planets than indicated by this figure. In fact, some large moons could be added to the list of potential candidates.

The parameter n_e suggests an integer, but as we have seen, it can also be interpreted as a probability when it is less than one.

7

Life and Exoplanets

7.1 The f_l Parameter in the Drake Equation

The fourth parameter is crucial: what is the fraction of habitable planets where some form of life will emerge? The search for exoplanets has identified some planets with high potential for harboring life. However, high potential does not necessarily imply the presence of life. This raises the question: what is the probability that habitable planets can host life? For instance, life can manifest in various forms, ranging from simple entities like bacteria, fungi, and viruses to far more complex organisms such as plants and animals. A possible answer to this question is providing an estimate of the parameter f_l of the Drake equation. The main impediment to estimating the fourth parameter stems from our current lack of understanding regarding the origin of life on Earth. As previously discussed, various theories, such as abiogenesis, panspermia, or a blend of both, have been proposed to explain this phenomenon. Let us introduce some simple chemistry concepts to delve further into this topic.

7.2 Organic Chemistry Compounds

Compounds derived directly or indirectly from living organisms, such as animals and plants, are termed organic compounds due to their carbon-based nature. However, it's worth noting that not all carbon-containing compounds originate from life, although the majority do. On the other hand, inorganic compounds are sourced from non-living materials, like rocks and minerals. In 1828, Friedrich Wöhler, a prominent German chemist, achieved a significant

© The Author(s), under exclusive license to Springer Nature Switzerland AG 2025
L. Vacca, *Life Beyond Earth*,
https://doi.org/10.1007/978-3-031-81695-6_7

milestone by synthesizing urea from ammonia cyanate [98]. This marked the first instance of an organic compound (urea) being synthesized from inorganic components. Urea, typically extracted from urine, serves various purposes, including industrial and agricultural uses. Another pivotal advancement in synthesizing organic compounds from inorganic sources was made by another German chemist, Hermann Kolbe, a key figure in organic chemistry. In 1845, Kolbe successfully synthesized acetic acid from carbon disulfide [99]. Acetic acid, found in various foods like apples, grapes, strawberries, and vinegar, plays a crucial role in lipid metabolism [100].

Despite Wöhler's synthesis, some chemists persisted in believing in vitalism, a theory positing the existence of a special vital force in organic materials. Kolbe, however, championed the notion that organic compounds could be synthesized through gradual processes. To illustrate this concept, he converted carbon disulfide into acetic acid over two years, starting in 1843. While these initial experiments demonstrated the potential for organic compounds to originate from inorganic sources, the ultimate challenge is the synthesis of life from inorganic matter.

To delve deeper into this complex issue, we turn to the renowned experiment conducted by Miller and Urey. Their objective was to recreate the primordial conditions hypothesized to have led to the emergence of life on Earth.

7.3   The Miller and Urey Experiment

In 1952, two American chemists, Stanley Miller, and Harold Urey, conducted a groundbreaking chemical experiment at the University of Chicago to reproduce the primordial terrestrial conditions that, according to the theory of abiogenesis, would have led to the appearance of the earliest forms of life on Earth [101].

The experiment used all the substances presumably present billions of years ago on Earth: water, methane, ammonia, and hydrogen. In addition, electric sparks were applied to a mixture of these substances in small vials, sealed for future studies, to simulate lightning most likely present before the appearance of life. In 2007, scientists who examined the sealed vials found that more than 20 amino acids were produced in this experiment. Amino acids are complex organic compounds that contain nitrogen, hydrogen, oxygen, and carbon, forming the basic building blocks of proteins. Protein synthesis uses only 20 amino acids [102] plus two non-standard amino acids: selenocysteine and pyrrolysine. Pyrrolysine is not present in humans. Thus, one could conclude that this experiment could finally prove the thesis of abiogenesis once and for all. However, the compounds produced by this experiment were found to be of racemic type. This result may constitute a difficulty for the theory of abiogenesis. What does "racemic" mean? To grasp the significance of this term, we need to delve into the concept of chirality [103].

7.4 Chirality

In organic chemistry, a chiral molecule cannot be aligned with its mirror image. An illustration of chirality is the distinction between the left and right hands, which cannot be perfectly overlapped. Chirality distinguishes one molecule from another, even if they have the same chemical composition-an observation made by Louis Pasteur and published in 1848 [104]. Essentially, a molecule's geometry is crucial in determining its function. A racemic compound contains equal amounts of molecules with two different chiralities: left-handed and right-handed. Unlike the compounds produced in the Miller and Urey experiment, life on Earth did not emerge in racemic form. Instead, its foundational components prefer one of the two chiral forms depending on the compound, a phenomenon known as homochirality. This asymmetry is essential for life as we know it. For instance, amino acids on Earth predominantly exist in the left-handed form (L-chirality), while those in the right-handed form (D-chirality) are typically harmful to living organisms. Conversely, all sugars crucial for life are predominantly right-handed (Fig. 7.1).

7.5 Chirality in Abiogenesis and Panspermia

What triggered the uniform chirality of biological compounds on Earth, paving the way for the earliest life forms approximately 4 billion years ago? Despite numerous abiogenetic theories proposing explanations for the emergence of homochirality on Earth, the underlying mechanism remains a mystery to date. All theories can be divided into two categories: (a) random occurrences that privilege one-handed form over the other and (b) a deterministic factor that gives rise to the separation of the two forms. In addition, some of these factors could be local to our planet or global throughout the universe. It is worth noting that D-chiral amino acids are often found as byproducts of living organisms. Indeed, several D-amino acids are present in small amounts in fruits, vegetables, and the milk of cows and goats [105]. D-chiral amino acids are components of many drugs, including bacterial antibiotics [106]. An intriguing discovery by Japanese scientist Noriko Fujii highlights that amino acids in

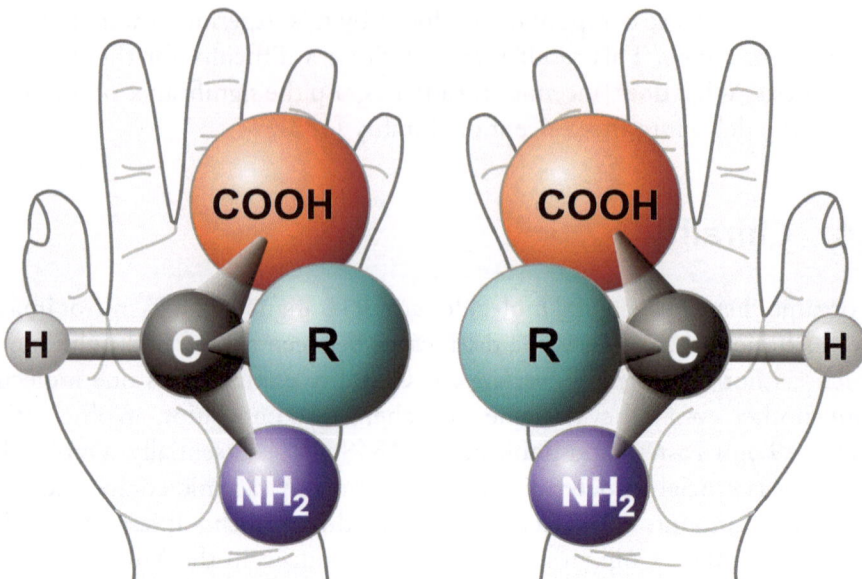

Fig. 7.1 Amino Acid Chirality With Hands from http://www.nai.arc.nasa.gov/. By Original: Unknown Vector. Chirality with hands.jpg, Public Domain, https://commons.wikimedia.org/w/index.php?curid=17071045

proteins within living cells do not maintain their homochirality consistently over their lifetimes [107]. Research has demonstrated that the racemization process of proteins is associated with lifespan duration, underscoring the significance of homochirality in all living organisms. One prominent hypothesis regarding homochirality suggests a potential asymmetry in the fundamental laws of nature as a contributing factor.

7.6   Parity Violation and Asymmetries in the Laws of Nature

A possible source of homochirality is based on the concept of parity violation in the weak force [108]. The weak force is one of the four fundamental forces of nature and is responsible for the radioactive decay of isotopes. Chinese-American physicist Chien-Shung-Wu observed the parity violation at Columbia University in 1956 [109]. Her experiment demonstrated that an electron emitted by Cobalt-60 tends to travel in the opposite direction of the spin of the nucleus. This symmetry violation produces a minimal energy difference between the ground states of the two types of chirality. This asymmetry in nature evokes a similar enigma: in theory, matter and antimatter should exist in equal quantities, yet matter's predominance over antimatter allowed the universe to exist. Therefore, it is plausible to speculate that minute asymmetries in the fundamental laws of nature may have played a role in the emergence of life in the universe, perhaps solely on Earth. Another hypothesis suggests that a chiral asymmetry emerged randomly, and the prevalent chirality subsequently dominated evolutionary processes. However, traces of the losing chirality have not been identified, suggesting this hypothesis is less likely.

Alternative theories, meanwhile, attribute homochirality to the Earth's rotation or magnetic field. The intensity of the Earth's magnetic field may not be sufficient to generate homochirality [111]. An alternative theory is that homochiral compounds may have been brought to Earth through meteorites after exposure to more powerful magnetic fields that stars and planets can generate. If true, this observation leads to consider an alternative theory to abiogenesis: panspermia.

7.7 The Panspermia Hypothesis: The Murchison Meteorite

The Murchison meteorite, which landed in 1969 in Murchison, Australia, is a carbon-rich meteorite weighing over 100 kg [112]. Due to its significant weight, it has been meticulously examined for its intriguing biological and chemical properties. A variety of amino acids were discovered on the meteorite, initially thought to be racemic. However, further investigation revealed that certain non-protein amino acids, including isovaline, were present in homochiral form [113]. This discovery appears to support the panspermia hypothesis, although the current level of meteorite activity is insufficient to create the homochirality imbalance necessary for the emergence of life on Earth.

Another potential explanation is the late heavy bombardment theory (LHB), which suggests a significant event approximately 4 billion years ago during the Neohadean and Eoarchean eras. This event involved many asteroids and comets colliding with Earth and other planets in the solar system. Evidence for this theory comes from lunar rocks collected from impact craters, which exhibit signs of rapid cataclysmic events resulting in rock melting. A possible cause for LHB is the migration of the giant planets. If true, the initial building blocks of life may have arrived on Earth during the late heavy bombardment through panspermia. These blocks may have found favorable conditions on Earth, leading to the formation of more complex molecules and the emergence of proto-life.

However, some scientists are skeptical of the panspermia theory, as it merely shifts focus to other celestial bodies where the early components of life may have originated.

7.8 Summary

Any estimate of the fraction of habitable planets with life emergence is predicated on understanding how primitive forms of life appeared on Earth. No definitive theory exists to explain the emergence of life on Earth approximately four billion years ago, shortly after the conditions for life were established. However, certain clues suggest that the panspermia hypothesis may have influenced the early appearance of life on Earth. These clues include:

- the inconclusive nature of the Miller-Urey experiment.
- the homochirality of life.
- the violation of parity and asymmetry in the laws of nature
- various meteorite findings like the Murchison meteorite.
- the speed at which early life formed 4 billion years ago.

Conversely, pure abiogenesis is supported by:

- the presence of a strong magnetic field on Earth.
- Earth sits in a habitable zone.
- Earth has a sizable atmosphere, land, and oceans with water.
- Miller-Urey experiment resulted in the formation of amino acids.

7.9 Microorganic Life May Be Not So Rare

Although we have yet to identify a definitive and scientifically proven mechanism responsible for the emergence of life on Earth nearly four billion years ago, we can still attempt to estimate the fourth parameter of the Drake equation using our limited understanding through a rough calculation. Within our solar system, Earth is the only known planet where conditions were suitable for the emergence of life. Furthermore, life formed as soon as the conditions were appropriate. This fact points to a non-random occurrence if one espouses abiogenesis. Some scientists have hypothesized that Venus or Mars may have harbored life in the distant past, although this remains speculative. Furthermore, it has been found that Mars's interior hosts ice and water oceans kilometers below its surface [115]. Mars explorations have not found any life on the red planet. However, one can speculate that some rudimental form of life may be swimming in one of the large subsurface oceans on Mars and the moons of the giant planets. Another important clue is life resilience, as the existence of extremophiles seems to suggest. In the next chapter, we will show that life on Earth has survived five major extinction events and several minor events. Given that Earth possessed the necessary conditions and life unequivocally thrived here as soon as it could, one could tentatively argue that life will arise in the case of planets and suns with the right conditions:

$$f_l \approx 1 \tag{7.1}$$

If we were to discover microorganisms inside the moons of the giant planets, such discovery should bolster our hypothesis that life arises where it finds a suitable environment.

8

Intelligent Life

The fifth parameter of the Drake equation is the probability that life will evolve into one or more intelligent species capable of communicating with other extraterrestrial species. Given that we have only observed intelligence on Earth, it's worth revisiting the unique sequence of events that led to humans' emergence. This sequence, filled with fascinating twists and turns, is a testament to the complexity and wonder of life's evolution.

8.1   The Path to *Homo Sapiens*

Life emerged on Earth approximately 3.8 billion years ago, in the form of prokaryotes 750 million years after the planet's formation [116]. Prokaryotes are primarily unicellular organisms, such as bacteria or archaea, characterized by the absence of a nucleus in their cells. Instead, their DNA floats freely within the cytoplasm, not enclosed by a membrane.

It's important to note that prokaryotes have consistently demonstrated remarkable survival skills, thriving in the challenging environments that existed on Earth billions of years ago. Their adaptability and resilience are truly impressive, and they have managed to survive and thrive in conditions that would be inhospitable to many other organisms. Their survivability is a function of their large degree of species diversification and their use of inorganic compounds to extract energy. Prokaryotes are ubiquitous on Earth, thriving in diverse environments ranging from the frigid glaciers of Antarctica to the ocean depths, volcanic environments, and beneath the soil. Around two billion years ago, during the Proterozoic Eon, eukaryotes emerged, representing a new type of organism characterized by cells containing a nucleus. One prevailing theory

© The Author(s), under exclusive license to Springer Nature Switzerland AG 2025   **61**
L. Vacca, *Life Beyond Earth*,
https://doi.org/10.1007/978-3-031-81695-6_8

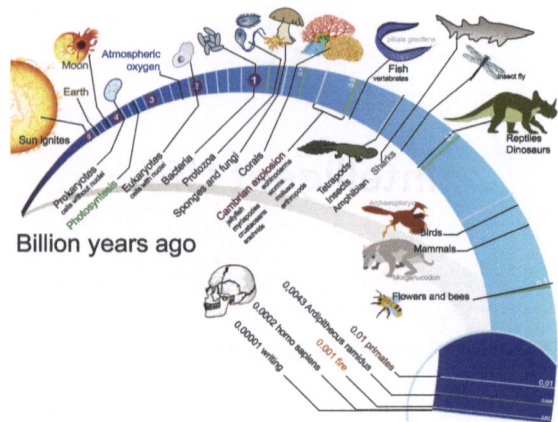

Fig. 8.1 Timeline Evolution of Life. By LadyofHats—Own work, CC0, https://commons.wikimedia.org/w/index.php?curid=20397531

explaining the origin of eukaryotic cells suggests that they arose through a symbiotic process among prokaryotic cells, a transformation that spanned over a billion years. Eukaryotic cells are significantly more sizeable and more complex than prokaryotic cells. In eukaryotic cells, there is a division of labor among various components: the nucleus houses the DNA, a membrane system facilitates the transport of nutrients, and mitochondria generate energy for the cell (Fig. 8.1).

Eukaryotic cells form the basis of plants, animals, and humans. However, the question remains: how did complex organisms like humans evolve from these relatively complex cells?

In 1858, English naturalists Charles Darwin and Alfred Russell Wallace put forth the theory of evolution as the primary source of the incredible life diversity we observe on Earth. Genetic mutations can arise from various factors, including damage to DNA, environmental influences, or errors during DNA replication. Through natural selection, mutations that enhance an individual's chances of survival and reproduction tend to be inherited by subsequent generations. The theory of evolution proposes the existence of a common ancestor known as LUCA (Last Universal Common Ancestor) from which all diverse forms of life on Earth descended, much like branches stemming from a tree [117]. Furthermore, the concept of LUCA is reinforced by the shared genetic code observed across different life forms. It should not be seen as a single life form but as a theoretical genetic formulation.

From the early eukaryotic cells and during the Proterozoic Eon [118], land fungi appeared 1.3 billion years ago [119], followed by the Cambrian explosion 541 million years ago [120]. As the name suggests, the Cambrian explosion is characterized by a tremendous acceleration in the diversification and complexity of marine life and animals for tens of millions of years. We now know that a remarkable diversification of life occurred thanks to the Burgess Shale [121] fossils in Canada's Yoho National Park. The diversification was marked by the emergence of mobile animals exhibiting evident anatomical complexity. Among the various species that flourished during this time were arthropods, algae, crustaceans, sponges, mollusks, worms, and chordates. Notably, chordates are mostly vertebrates, one of our distant ancestors (Fig. 8.2).

In a significant discovery, the fossil of a minuscule creature dating back to the Cambrian explosion was unearthed in the Burgess Shale fossil beds in 1911 by Charles Doolittle Walcott. This creature, known as *Pikaia gracilens*, has been classified as a chordate by researchers from the University of Cambridge, University of Toronto, and the Royal Ontario Museum [122]. Chordates are the group from which vertebrates, including mammals, originated. Mammals, fish, birds, and reptiles share a common ancestry with Pikaia. For instance, Morganucodon, tiny rodents, are very early mammals that appeared roughly 200 million years ago, coexisting with the dinosaurs on Earth [123].

Fig. 8.2 Model of *Homo erectus* man in The Natural History Museum, Vienna By Jakub Hałun, CC BY-SA 4.0, https://commons.wikimedia.org/w/index.php?curid=113008349

The evolutionary path diverged further as primates branched out of other mammals. Fossil studies show that archaic primates appeared between 55 and 65 million years ago [124]. Apes and Old World monkeys split around 30 million years ago [125]. Apes are characterized by their larger size, absence of a tail, and increased brain size. The ape lineage further split into two branches: lesser apes, such as gibbons, and great apes, which include gorillas, chimpanzees, and bonobos [126]. Humans diverged from the great apes less than 6 million years ago [127].

One notable early human species is *Homo erectus*, a hominid capable of walking upright, which emerged in the southern part of Africa and later migrated to Asia starting around 2 million years ago [128]. The evolutionary journey leading to the appearance of early humans spans a vast timescale. Notably, this path might have been changed if not for the extinction of dinosaurs caused by a catastrophic asteroid impact event. Hence, extinction events are crucial in shaping life's trajectory on Earth. Now, let us briefly explore the history of extinction events to gain insight into how we arrived at our present state.

8.2   The History of Life Extinctions Events on Earth

Once life appears on a planet or moon, it becomes incredibly resilient. To underscore this resilience, let us explore the life extinction events that have unfolded on Earth since the emergence of the earliest life forms approximately 3.7 billion years ago.

These events could have potentially extinguished life as we know it on Earth. Fortunately for us, they did not succeed. Nonetheless, they led to rapid and substantial reductions in biodiversity across all levels of the Earth's ecosystems.

Many of these extinction events significantly impeded the evolutionary progress of life on Earth by millions of years. In some instances, they even resulted in a complete redirection of the evolutionary trajectory, as exemplified by the demise of the dinosaurs.

According to a study published by American paleontologists David Raup and Jack Sepkoski in 1982 [129], there have been at least five major life extinction events in the last 600 million years, which have led, in some cases, to an extinction rate over 80% on Earth. These are, in chronological order:

- The Ordovician-Silurian Extinction occurred from 460 to 435 million years ago [130].
- The Devonian Extinction occurred from 383 to 359 million years ago [131].

- the Permian-Triassic Extinction, which occurred about 252 million years ago.
- the Triassic-Jurassic Extinction, which occurred about 200 million years ago [132].
- and finally, the Cretaceous-Tertiary Extinction, which occurred 66 million Years Ago.

Let us review the causes and the impact of each extinction event one by one.

8.2.1 The Ordovician-Silurian Extinction

This extinction event ranks second most severe among the five major events, surpassed only by the Permian-Triassic event. During the Ordovician period, early marine invertebrates dominated the seas. Algae, sponges, early fish species, snails, cephalopods, corals, crinoids, and gastropods thrived abundantly. On land, primitive plants were prevalent, and moist and warm conditions characterized the climate [133]. Most of the landmasses were consolidated into a supercontinent known as Gondwana, which gradually migrated toward the South Pole. The term "Ordovician" originates from the name of an ancient Celtic tribe renowned for their warfare, the Ordovices, chosen by English geologist Charles Lapworth in 1879. The extinction event unfolded in multiple phases. First, A period of glaciation known as LOMEI-1 occurred, followed by a subsequent drop in sea levels. These environmental changes posed challenges for many species without tolerance to such abrupt temperature fluctuations. The glaciation period was characterized by the emergence of new species adapted to the colder conditions. Subsequently, a second phase of extinction, known as the LOMEI-2, occurred as the glaciers receded and temperatures warmed [134]. This phase was characterized by oxygen depletion and a significant increase in toxic sulfide production—more than 80% of all marine species perished during this tumultuous period. Notably, the ecosystem was able to bounce back to the conditions preceding this event in a relatively short period of ten million years [135].

Possible causes of this extinction event were:

- True polar wander. A migration of a planet's body relative to its spin axis [130]
- Volcanism [136].
- A gamma-ray burst may have been the cause, as advanced by science writer Philip Ball and others [137, 138].

The first two hypotheses are the most plausible. However improbable, let us delve into what a gamma-ray burst entails and its potential effects on a planet or moon.

Gamma-Ray Bursts and Life

Gamma-ray bursts are a cosmic threat to life. When a star undergoes a supernova explosion, or two neutron stars collide and merge, they may unleash the most intense beam of electromagnetic energy known in the universe: a gamma-ray burst. Gamma rays, the most energetic form of electromagnetic radiation, pose a more significant threat to life than X-rays commonly used in diagnostic medicine. These bursts can persist from a few seconds to several hours, emitting more energy than the Sun over its entire lifespan. Traveling billions of light-years, a gamma-ray burst striking a planet can strip it entirely of its ozone layer or even its entire atmosphere, owing to its immense energy. For instance, a gamma-ray burst emitted at a distance of 10,000 light-years could seriously threaten life. Consequently, such an event can have galactic-scale repercussions for life. Cosmic phenomena like gamma-ray bursts play a significant role in the birth and demise of life at planetary, systemic, and even galactic levels. In 2022, a gamma-ray burst was detected by the European Space Agency, lasting approximately 10 min [139]. Fortunately, this burst originated from a galaxy two billion light-years away and lacked the energy to damage our atmosphere. Typically, gamma-ray bursts are detected less often than supernovae. However, that does not imply that they are less dangerous to life.

8.2.2   The Devonian Extinction

Termed the Late Devonian Extinction, this event occurred towards the conclusion of the Devonian period. Extinction occurred as a prolonged series of events. The Devonian era marked the emergence of land animals, tetrapods, and arthropods, encompassing four-legged vertebrates, insects, and spiders [140]. New, more evolved types of fish were present.

During the Late Devonian period, terrestrial plants evolved roots, leaves, and seeds, increasing to considerable sizes. This botanical diversification is often referred to as the "Devonian Explosion." Geographically, the Devonian era featured the presence of two major landmasses: Gondwana and Laurussia [141].

Towards the end of the Devonian period, a prolonged mass extinction event unfolded over tens of millions of years as a series of pulses. Mirroring previous events, an extended climate instability followed due to carbon dioxide depletion. Consequently, it was followed by a period of glaciation, coinciding with diminished sea levels and scarcity of oxygen in the oceans (anoxia). Consequently, numerous marine vertebrates migrated to terrestrial habitats. Species populations plummeted by an estimated 75% [142]. Several potential causes for this extinction event include:

- Volcanism. Strong volcanic activity.
- Expansion of land plants reducing carbon dioxide [140].
- Asteroid or comet events [143].
- A supernova event [144].

As we have established with the infamous Chichxulub asteroid, celestial bodies like asteroids and meteorites can potentially trigger extinction events. Supernovae also possess this capacity. When located within approximately 50–300 light-years of Earth, they can emit X-rays and gamma rays that might erode the ozone layer, shielding us from harmful ultraviolet radiation and leading to catastrophic consequences for marine reef ecosystems. Nearby planets and moons could face the risk of their oceans boiling away. Determining a safe distance from a supernova remains uncertain; it might require several hundreds of light years. Debate persists regarding the frequency of supernovae that pose a threat, occurring perhaps every few tens of millions to hundreds of millions of years. Many scientists concur that supernovae may have instigated extinction events in the past.

8.2.3 The Permian-Triassic Extinction

The Permian-Triassic extinction, occurring approximately 252 million years ago, represents a succession of events near the conclusion of the Permian period. This series of events resulted in the most extensive mass extinction among Earth's five major extinction events. It is also known as the "Great Dying" to stress its catastrophic level of life loss. Over 80% of all marine species and more than 70% of all land species perished during this series of events [145], which probably lasted about 50–60 thousand years [146].

During the onset of the Permian period, Pangaea, a new supercontinent, emerged through the amalgamation of the original landmasses Gondwana and Laurussia. Alongside Pangaea, the Panthalassic Ocean, a vast superocean, also took shape and eventually broke up at the beginning of the Triassic period. Reef ecosystems, sponges, corals, and ammonites thrived during this period, alongside the emergence of bony fish resembling modern species. On land, mosses gave way to gymnosperms, which include conifers among their ranks. Insects continued their proliferation, while synapsids and sauropsids dominated terrestrial ecosystems [147, 148]. Synapsids, precursors to mammals, faced severe decline following the Permian extinction, whereas sauropsids, which evolved into dinosaurs, reptiles, and birds, fared comparatively better, ushering in the age of dinosaurs.

The cause of the Permian-Triassic mass extinction remains unclear, with potential factors including:

- A series of volcanic eruptions on a massive scale in Siberia and China [149].
- The impact of a large asteroid in an area of what is today Australia [150].
- Decrease in oceanic oxygen levels (anoxia) [150].
- Lowering of the sea levels [150].
- Changes in oceanic chemistry, specifically due to increased carbon dioxide [150].

The most plausible explanation for the Permian-Triassic Extinction is attributed to a sequence of volcanic eruptions triggering what is known as a "volcanic winter." This phenomenon arises when extensive volcanic ash, laden with sulfur dioxide and carbon dioxide, is ejected into the stratosphere. These particles obscure sunlight and reflect solar radiation, resulting in surface cooling on Earth. This cooling effect can induce prolonged periods of glaciation, causing severe repercussions for life forms.

8.2.4   The Triassic-Jurassic Extinction

As in the previous extinction, tremendous volcanic activity in Central Atlantic Magmatic Province generated global warming 201 million years ago, elevating the carbon dioxide concentration levels in the atmosphere and leading to ocean acidification [151]. Due to this volcanic activity, more than 70% of all land and marine species perished [153]. This event mainly killed corals, amphibians, and reptile species while mostly sparing plants and dinosaurs [132]. This event made dinosaurs the ruling animal species on Earth for over 100 million years until the last of the five extinction events.

8.2.5   The Cretaceous-Tertiary Extinction

The Cretaceous-Tertiary extinction is the final of the five major extinctions responsible for more than 70% of all plant and animal species [154]. This extinction was mainly attributed to the impact of a massive asteroid in the Yucatan peninsula of the Gulf of Mexico. This theory was initially advanced by American scientist Luis Alvarez et al. [155] in the early eighties and is widely accepted.

As mentioned earlier, the asteroid impact, an asteroid with a size comparable to a small city, created the Chicxulub crater. This event ejected a vast amount of material into the atmosphere as aerosols. These aerosols blocked sunlight and hindered photosynthesis in plants, leading to the decline of herbivorous dinosaurs reliant on vegetation. Consequently, carnivorous dinosaurs faced extinction due to the reduction in herbivore populations. Iconic species like the Tyrannosaurus Rex and the Spinosaurus vanished entirely. Many species were wiped out, including pterosaurs, mosasaurs, plesiosaurs, ichthyosaurs, ammonites, and cephalopods. However, birds, snakes, lizards, and mammals managed to survive. Mammals, in particular, thrived by feeding on insects and aquatic plants.

Following this extinction event, life flourished once again on Earth approximately between 4 and 10 million years after the demise of the dinosaurs [156].

8.3 The Lesson from Extinction Events

While there is no certainty regarding the existence of extraterrestrial life, if it does exist, it will probably share biological similarities with life on Earth. This allows us to draw some extrapolations from studying the evolution of life on our planet. These considerations include:

- Lifeforms that exhibit characteristics more adapt to the environment are more likely to reproduce and thrive.
- Once life appears on a planet or moon, it shows incredible resilience even to globally adverse events.
- Planets and moons will be subject to extinction events with varying severity.
- Some extinction events may be due to factors specific to a planet, for instance, volcanism.
- Other extinction events could be coming from space, for instance, solar flares, gamma-ray bursts, and supernovae.
- Extinction events that are survivable have impacted the evolution of the species.

Life has demonstrated remarkable resilience on Earth, rebounding from major extinction events at least five times. Around 18 minor extinction events have occurred approximately every 30 million years [157]. Despite these catastrophic events, life has persisted in diverse and extreme environments. Organisms thrive in the coldest polar regions, the hottest deserts like Death Valley, the deepest ocean trenches such as the Mariana Trench, and even in extreme pressure and temperature conditions deep beneath the Earth's surface. Hypothesizing on the potential life trajectory if dinosaurs had not gone extinct is uncertain. It's possible that a subspecies of dinosaurs with increased brain size and the capacity for tool use could have emerged as the dominant species. However, just as on Earth, extinction events will likely occur on any planet or moon hosting life over sufficiently long periods.

At minimum, such exobodies will be exposed to cosmic events such as asteroid collisions, gamma-ray bursts, and supernovae. One may expect that when conditions are conducive to survival, such events may exert an evolutionary selection that ultimately favors the emergence of intelligent species. What aliens may look like? Should we expect to be like us or unimaginably different?

8.4   Comparative Biology

No one knows the answers to these questions, as no sample of alien life is available to us. However, Dr. Arik Kershenbaum, a zoologist at the University of Cambridge, has tackled these questions by writing a groundbreaking book in 2020 that draws from his experience as a zoologist. His book is titled "The Zoologist's Guide to the Galaxy" [158]. Dr. Kershenbaum, whose primary interest is animal communication, states it is hard to say what aliens look like, but we can infer the functions that they may have by studying life on Earth. The laws of physics and chemistry apply everywhere in the universe. Hence, one should expect the laws of biology and, more specifically, the law of natural selection to be universal. It is natural to expect that life on Earth and aliens have plenty of functions in common. Dr. Kershenbaum recognizes a level of complexity in biology that may yield unpredictable effects, but the general principles still apply. What should we expect from alien life, according to Dr. Kershenbaum? Aliens should evolve according to natural selection, irrespective of their biological makeup. Life doesn't need to be based on DNA to evolve. Six fundamental features should be subject to evolution: movement, communication, intelligence, sociality and cooperation, information, and language. Dr. Kershenbaum also considers the possibility of AI-based life that passes knowledge from generation to generation as humans do. Dr. Kershenbaum is a proponent of convergent evolution. He cites the example of flying birds and insects, focusing on the usefulness of flight for such creatures. While birds and insects are very distant species, they developed the ability to fly because it conferred them an evolutionary advantage.

8.4.1   Aliens and Machines

In the last 60 years, with the advent of neural networks and powerful computers, humans have created intelligent tools that have generating and predictive power. More recently, people have started wearing wearables to stay informed about the world, communicate, and monitor their bodies and habits. At some point, all technological wearables will carry AI and nanoprobes capable of repairing damage in the human body. This trend will accelerate in the coming years. It is easy to imagine that humans will eventually merge with intelligent machines, resulting in an enhanced life form capable of performing daily tasks faster and more accurately. Presumably, intelligent extraterrestrial species have already merged with intelligent machines due to their great benefits. Hence, such species may look like a machine on the surface but have organic matter inside the machine. Therefore, we should not be surprised if aliens will look very different from us. The cyber aliens may decide to shed organic material components in their bodies altogether. For instance, they could upload their cognitive functions into quantum computers, bypassing the evolutionary trait that makes them mortal and gaining almost unlimited knowledge.

8.5   Conclusions

At present, no one knows the probability that life may evolve into an intelligent species like *Homo sapiens* on an exobody. In fact, throughout its 3.7 billion years on Earth, life could have faced complete extinction at any moment. This event cannot be ruled out a priori. Is life merely a stroke of luck that has persisted through such challenges, even flourishing, or does life possess an extraordinary capacity to adapt to even the most extreme environmental circumstances and events? Drawing from the history of our planet, it seems that life possesses a remarkable ability to thrive in the face of adversity. If life is resilient, comparative biology teaches us that intelligence is one of the byproducts of the law of natural selection, and one should expect it to appear in extraterrestrial biology. However, notwithstanding the previous qualitative considerations, no estimate can be reliably provided for f_i.

$$\boxed{f_i = ?}$$
(8.1)

The reader should not be surprised that we are not producing an estimate for the intelligence parameter. Our approach is to treat this parameter instead as an unknown. Later, when we draw some conclusions on the Drake equation, we consider different cases for this parameter and their impact on the number of communicating extraterrestrial civilizations.

9

Communication

9.1   The f_c Parameter in the Drake's Equation

The sixth parameter of the Drake equation involves an estimation of the fraction of intelligent civilizations that have proceeded in the past to communicate with other species, probably broadcasting an electromagnetic signal through space. Why does the signal need to be electromagnetic?

Electromagnetic signals are favored because they travel at the speed of light, making them the swiftest means of disseminating information. With numerous advanced telescope facilities across the globe capable of capturing radio signals from space, radio signals are the preferred option due to their broad frequency range, which spans from 20 kHz to 300 GHz, encompassing a significant portion of the electromagnetic spectrum suitable for both sending and receiving information. In addition, they travel through interstellar dust and are less likely to be absorbed or scattered than their higher-frequency counterparts. Higher frequencies pose challenges due to their heightened energy demands and susceptibility to interference from natural sources like stars emitting infrared and visible light, as well as the harmful effects of ultraviolet, X-rays, and gamma rays, which, in turn, merits their exclusion. In addition, radio signals may encounter lots of interference due to the various solar flares along their trajectories. Hence, even radio signals are far from being a perfect means of communication. Another possibility is that aliens can use laser signals for communication. Lasers are narrow, coherent beams of light.

Having stated this, what technological level would an extraterrestrial civilization need to communicate with other civilizations?

© The Author(s), under exclusive license to Springer Nature Switzerland AG 2025
L. Vacca, *Life Beyond Earth*,
https://doi.org/10.1007/978-3-031-81695-6_9

9.2   The Kardashev Scale

In 1964, Soviet astronomer Nikolai Kardashev proposed a scale to gauge the technological advancement of extraterrestrial civilizations based on their energy consumption [159]. This scale draws on historical trends, from our ancient ancestors harnessing heat through combustion to our modern mastery of nuclear energy. Presently, our energy production dwarfs that of our ancestors, albeit a significant portion still stems from burning fossil fuels. Notably, a clear link exists between a nation's standard of living and its energy production and consumption; more technologically advanced nations tend to produce and consume more energy. It is enough to consider our need for energy for air conditioning and its impact on our civilization. Through energy production, societies facilitate vast information exchange, expedite long-distance travel, and regulate home temperatures. While other factors contribute to a higher standard of living, Kardashev's framework focuses on energy production and utilization.

Kardashev delineated three types of civilizations based on their energy-harnessing capabilities. A Type I civilization can tap into all the energy resources on their planet. In contrast, a Type II can harness the energy of their sun, and a Type III can harness the energy of an entire galaxy. A Type I civilization can generate power equivalent to the combined radiation from their sun, wind, and geothermal sources, estimated at $10**16$ watts by Carl Sagan [160], surpassing our current power production by several orders of magnitude. A Type I civilization can also control natural hazards like weather phenomena, earthquakes, and volcanic eruptions.

Are we a Type I civilization? On a logarithmic scale, we are about a Type 0.7 civilization. According to British-American physicist Freeman Dyson, who first conceived the possible existence of the Dyson sphere, estimated that, within 200 years or so, humanity should reach Type I status [161]. A type II civilization can extract all the energy their sun emits using a *Dyson sphere*, a concept first conceived by the physicist Freeman Dyson. Such a sphere could be a collection of probes that capture light from the sun in large quantities. A Dyson sphere could also denote a gigantic shell around a sun capable of absorbing its entire radiative energy. Such a shell must be positioned at a radius larger than our distance from the Sun to avoid heavy damage from the solar winds (Fig. 9.1).

Finally, a type III civilization could have the capability to harness the energy of hundreds of billions of stars within their galaxy. Consequently, their power output would exceed the output of a Type II civilization by tens of billions of times. The disparity in power output between each type of civilization spans at least ten orders of magnitude. Notably, Type II and III civilizations could

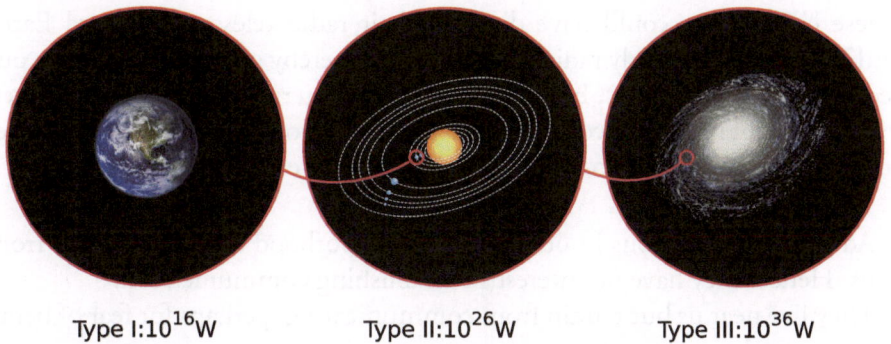

Type I:10^{16}W   Type II:10^{26}W   Type III:10^{36}W

Fig. 9.1 Kardashev scale. By Indif.:1 Earth.png4 Milky Way.png, CC BY-SA 3.0, https:// commons.wikimedia.org/w/index.php?curid=29015315

transmit signals in all directions (isotropic) with such intensity that they could be detected by intelligent species millions of light-years away. This underscores the vast technological prowess and reach of such advanced civilizations.

Given this technological assumption, why haven't we received a signal from space generated by an intelligent source?

9.3   The Great Silence

Despite the abundance of listening radio telescopes, humanity has not yet intercepted messages from advanced extraterrestrial civilizations. One notable exception is the brief "Wow!" signal detected in 1977 by the Ohio State University radio telescope. However, this raises an important question: Is it realistic to assume that an advanced alien species would even view us as potential interlocutors? Humanity acquired the technological capability to capture and potentially interpret extraterrestrial messages only within the last 100 years or so of our existence on Earth.

No alien civilization farther from us than 100 light-years have, *a priori*, detected our level of technology from our radio transmissions. To put into perspective, the famous Italian inventor Guglielmo Marconi invented radio transmission in 1895 by sending a signal from his father's estate more than three kilometers away and over the hills. The first commercial radio station started broadcasting in 1920. The station's name was KDKA, based in Wilkinsburg, Pennsylvania. A tremendous growth in radio broadcasting occurred between 1920 and 1950.

Another hypothesis posits that advanced civilizations may live in a system near our Solar System, perhaps within a few tens of light-years from us.

These civilizations could have directed their radio telescopes toward Earth and intercepted our early radio transmissions, thereby gaining insight into our technological capabilities. However, even if we entertain the idea of advanced extraterrestrial life in our cosmic vicinity, the question remains: why are we yet to hear anything? Let us consider several potential explanations:

- Advanced civilizations in our galactic neighborhood are too different from us. Hence, they have no interest in establishing communication.
- They live near us but refrain from communicating, perhaps for fear of being intercepted.
- They have sent us a signal, but we have not been able to either capture it or recognize it as such.
- They are waiting until we become a more mature civilization.
- Communication is targeted only to certain types of civilizations.
- Life nearby is too primitive for communication. Advanced civilizations are situated too far away.
- They have not heard us.

For numerous reasons, private communication within our society is significantly safer and more targeted. We choose not to trumpet what we consider private to everyone, risking unintended listeners picking it up. To underscore this point, one can observe the vital role of encryption in cellphone and computer communications. Thus, it could be posited that communication is most desirable when it's confined to the involved parties. This could be achieved through cloaked signals with highly directional properties. This might help explain, at least partially, the profound silence. However, even a cloaked signal carries the risk of interception and decryption by a third party, indicating that all forms of communication entail some risk. What might pique the interest of an alien civilization and prompt them to initiate communication? For example, an alien civilization detecting biosignatures on our planet from millions of light-years away might opt to transmit a signal to us. The sending species may anticipate that intelligent beings already inhabit the earth or at least one intelligent species will evolve here. But what exactly constitutes a biosignature, and how do we search for them?

9.4   Biosignatures Outside the Solar System

A biosignature refers to a biological or chemical substance, often in gaseous form, that could indicate the presence of life. The most effective method for searching for life involves sending probes to planets, moons, and even asteroids. However, given our current technological capabilities, such exploration is mainly confined to our solar system. For instance, it would take many thousand years to travel to Proxima Centauri b, four light years away from us, even with our fastest spaceships.

Luckily, it is possible to search for life on exobodies using spectroscopy, at least in principle. As it is well known from elementary physics, the light emitted by a star is characterized by a frequency spectrum, which is the amount of energy content present in the emitted radiation at a given frequency. The light emitted by a star can undergo partial absorption and scattering by the chemical substances within the atmosphere of a neighboring planet. Differences become apparent when comparing the frequency spectrum of unaltered starlight with that passed through a planet's atmosphere. These differences manifest as spectral lines within a narrow frequency range, indicating the presence of specific elements or compounds within the planet's atmosphere. For example, space telescopes such as the James Webb Space Telescope are equipped to capture and dissect infrared light across multiple frequencies. Additionally, ground-based telescopes can analyze light spectra from planets like ours. Generally, rocky planets with atmospheres located within the habitable zone of their parent star are considered potential candidates for spectroscopic investigation.

But what elements are considered biosignatures? Water is undeniably crucial for life, based on our understanding of life on Earth. Comprised of hydrogen and oxygen, it stands as one of the most abundant compounds in the universe, along with carbon monoxide, molecular hydrogen, hydroxyl radical, methane, and ammonium. Hydrogen is the most abundant element, followed by helium, with oxygen ranking third. Additionally, carbon, another prevalent element, serves as a significant biosignature. Compounds also play a vital role as biosignatures, such as water, molecular oxygen O_2, carbon dioxide CO_2, nitrous oxide N_2O, molecular nitrogen N_2, methane CH_4, methyl chloride CH_3Cl and many others [162].

It is important to remark that, although promising, the search for biosignatures is only at a nascent stage.

9.5   The Von Neumann Probes

We have already pointed out that, at this stage, direct exoplanet exploration would take an enormous amount of time, even using our fastest types of propulsion. However, there is an alternative way of exploring a galaxy. Enter the idea of the Von Neumann probe, as the Hungarian mathematician and physicist John Von Neumann proposed. In a nutshell, a Von Neumann probe is a probe that travels to a planet, moon, or asteroid and mines them for raw material to build a copy of itself. Then, the old and new probes move to other planets to repeat the same job. It is easy to imagine that the probes will grow exponentially and spread throughout the galaxy. Before long, a rapid proliferation of probes spans the entire galaxy. A highly developed civilization capable of constructing self-replicating probes could systematically explore the Milky Way in a much shorter time than it takes to traverse it. Identifying a Von Neumann probe would serve as a techno signature, offering evidence supporting the existence of an advanced entity or civilization accountable for its creation. What are some other possible technosignatures?

- Spaceships capable of flying at speeds that allow interstellar travel.
- Dyson spheres or other objects that can generate planetary or star-level amounts of energy.
- Any space objects that don't obey the natural motion under the influence of gravitational attraction.
- Large objects whose orbits are modified by advanced technology, thus seeming to violate the law of gravity.
- Any cosmic characteristics that physical laws cannot explain.

For instance, the star KIC 8462852, also known as Tabby's star in tribute to the American astronomer Tabetha Boyajian, has displayed significant fluctuations in its brightness since its observation started in 2015 [163]. These fluctuations prompted some astronomers to propose the presence of a Dyson sphere encircling the star. However, subsequent research has suggested that the phenomenon is likely caused by an unusually high amount of dust surrounding the star, possibly resulting from the destruction of an exomoon [164]. Nevertheless, the potential for large technological structures to interact with stars and planets remains a hallmark of intelligent activity.

9.6   Estimates

One could receive two types of messages from space: intentional and unintentional. We expect unintentional messages to be emitted virtually by every advanced civilization during their life, most likely for limited periods. Over the past 100 years, humanity has unintentionally sent electromagnetic waves into space in all directions. Our capturing intelligent unintentional messages could be compared to the event of a person hearing music played extremely loud by a passerby. For this reason, our estimate of the f_c parameter is of the order of unity in the case of unintentional messages:

$$f_{c_{unintentional}} \approx 1 \qquad (9.1)$$

The trouble, however, with our unintentional messages is that they are typically too faint. For instance, our early and current radio emissions were not generated with enough power for space communication over light-year distances. The total energy required for such emissions depends on the distance to the target and follows the inverse square law, meaning they significantly weaken with distance. However, a more advanced civilization with enormous energy could transmit an isotropic intentional signal, hoping to be heard by others. For instance, a civilization that faces a cosmic danger could send a powerful signal directed at other civilizations of an impending threat.

The likelihood of intentional messages coming solely in our direction should be pretty low. Firstly, we haven't even reached Type I civilization status; if there are other civilizations, we are probably an infant civilization from a technological standpoint. The capture and deciphering of intentional messages could require more advanced technology than we currently have. Deliberate and unintentional messages could reach us undetected due to our scientific and technological limitations. Moreover, given the technological gap, a Type II or III civilization capable of transmitting such signals would learn little from us, scientifically speaking. Studying us from afar is such civilizations' most convenient course of action. Consequently, the number of advanced civilizations attempting to contact us directly should be close to zero.

$$f_{c_{intentional}} \approx 0 \qquad (9.2)$$

10

Alien Transmission Duration

10.1 Communication and Civilization Lifespan

The last parameter of the Drake equation is L, the length of time during which an extraterrestrial civilization sends signals into space. Estimating this parameter proves highly challenging. Firstly, it is not sure that the duration of communication correlates with the lifespan of a civilization. As previously noted, humanity has only been emitting signals into space for the past century, despite *Homo sapiens* existing for several hundreds of thousands of years. While an alien civilization may strive to communicate, more advanced counterparts may supplant them. History provides examples; tens of thousands of years ago, *Homo sapiens* and Neanderthals competed for resources, with some theories suggesting that this competition hastened the Neanderthals' demise. Other factors that may have hastened the Neanderthals' demise include climate change and disease. Moreover, in many science fiction narratives, robots and artificial intelligence ultimately replace humans as the dominant civilization on Earth in the distant future. Some futurists have foretold the advent of "Homo technologicus". Likely, humans will fully merge with AI technology in the future. Regardless of the motivations behind communicating with distant extraterrestrial civilizations, these reasons will inevitably undergo scrutiny and reassessment, irrespective of the civilization's capacity to initiate and maintain communication. Furthermore, even advanced civilizations can encounter demise. They could perish due to internal factors such as nuclear wars, volcanic eruptions, runaway global warming, and so on. Alternatively, they could perish due to some cosmic extinction event. A civilization facing an imminent extinction event might endeavor to preserve its knowledge and culture by sharing it with other civilizations, akin to how we store literature, art, and scientific

L. Vacca, *Life Beyond Earth*,
https://doi.org/10.1007/978-3-031-81695-6_10

Fig. 10.1 Karl G. Jansky Very Large Array. By Mihaisiscanu, CC BY-SA 4.0, https://commons.wikimedia.org/w/index.php?curid=49889418

discoveries in books and digital repositories for future generations. There is an incentive for an advanced civilization near its demise to share its knowledge and culture with other civilizations, hoping its history will be preserved (Fig. 10.1).

10.2   Communication is Knowledge Acquisition

In any case, the presence of life on an extraterrestrial body would remain detectable to an advanced civilization seeking another extraterrestrial life, thanks to its biosignatures and, if present, its technosignatures. Undoubtedly, the findings of biosignatures and technosignatures constitute a discovery of the highest importance.

However, such detection may not warrant any further efforts from a cultural exchange standpoint when dealing solely with primitive life forms such as bacteria. There is an ethical argument for not interfering with the evolution of less advanced life forms. There could also be an argument that when two completely different life forms meet, the encounter may be dangerous for both. Conversely, an alien civilization could make a tremendous and lengthy effort to establish a line of communication to enhance their survivability through knowledge or help from a more advanced species. This endeavor could persist for as long as the civilization continues to seek help. This effort may either (A) conclude without achieving its goal or (B) result in the civilization achieving self-sufficiency, rendering external assistance unnecessary. What is the typical lifespan of a civilization before facing extinction?

The American astrophysicist J. Richard Gott developed an argument focused on humanity's survival time, hence its name, "the doomsday argument."

10.3 The Doomsday Argument

Gott's argument is premised on the Copernican principle that nothing is unique about us nor the time and place we live in [165]. Gott assumes that the fact we have survived until now may be correlated with our future survival. A typical argument follows: humankind was born at time $t_{beginning}$ and will end at time t_{end}. A human observer happens to be living between these two times at a random time t_{now} with a uniform distribution. The uniform distribution is assumed *a priori* to a lack of knowledge regarding the actual distribution. Based on such assumption, we introduce the quantity δ, a randomly generated number ranging from 0 to 1 as follows:

$$\delta = \frac{t_{now} - t_{beginning}}{t_{end} - t_{beginning}} \tag{10.1}$$

Hence, there is a 95% probability that δ falls within a range of 0.025 to 0.975.

Simple algebra shows that the time $t_{remaining}$ left until t_{end} is related to the time we have survived $t_{survival} = t_{now} - t_{beginning}$ by the following formula:

$$\frac{t_{survival}}{39} < t_{remaining} < 39 * t_{survival} \tag{10.2}$$

The range is predicated on a 95% confidence level.

For instance, *Homo sapiens* has been around for about 300 hundred thousand years [166]. A simple calculation yields that the human species is likely to be around at least for $300,000/39 \approx 7692$ years and perish within $300,000 * 39 = 11.7$ million years.

Gott's argument relies on likelihood rather than certainty. While many counterarguments can be raised against it, such as the possibility that humanity may not face an inevitable end driven by uniform distribution, there may still be some truth to Gott's argument. As civilizations age, they tend to become more technologically advanced, theoretically increasing their ability to prolong their existence. There could be underlying, unknown factors contributing to their longevity.

Conversely, technological advancements themselves may pose existential risks if misused. Examples are nuclear weapons, biotechnology, and AI. Therefore, there may be a balance between the benefits and the risks that new technologies may offer, which may entail a more thorough analysis of our species's survival time.

10.4 Space Colonization

If Gott's argument holds any validity, any alien civilization much older than us should be expected to be around for a long while. Given their presumed highly advanced technological level, it is unlikely that such civilizations may be interested in exchanging information with us. Moreover, even if they attempted to signal us, it might be futile considering the immense technological disparity between us and them. For instance, consider Leonardo da Vinci, a painter, engineer, and polymath of extraordinary talent who lived in Italy during the High Renaissance. Despite his unquestionable brilliance, imagine if he were given a copy of a microchip design from the future. Extracting useful information from such a design would be practically impossible even for a genius of his caliber, especially with a technological gap spanning 500 years. Now, let us switch to an advanced alien civilization millions of years ahead. We would have no chance to make sense of the simplest device created by such a civilization. Their understanding of physics may be completely different from ours. And they may not even use mathematics as we know it.

The famous science fiction writer Arthur C. Clarke, the author of the bestseller novel "2001: A Space Odyssey", once wrote that "any sufficiently advanced technology is indistinguishable from magic." Any advanced alien technology would probably appear magical to us.

Instead of broadcasting signals into space, an advanced alien species might be inclined to explore space and establish colonies on new celestial bodies. Drawing a parallel with human history, *Homo sapiens* departed from their African homeland around 180,000 years ago, migrating to Europe, Asia, and the Middle East, likely driven by climatic changes and ensuing droughts [167]. This migration was facilitated by the development of new tools and the mastery of farming techniques. Similarly, alien species might opt to settle new worlds once local resources are depleted, their parent star nears the end of its life cycle, or to evade cosmic threats. The underlying motive is survival. A civilization dispersed across multiple solar systems stands a greater chance of long-term survival than one confined to a single solar system. Another possibility is that alien species could use rogue planets to traverse between solar systems, as researcher

Irina Romanovskaya suggested [168]. Type II civilizations might even manipulate the orbits of their planets to seek out new systems or to move inside a habitable zone. Alternatively, they could attach propulsion systems to asteroids for interstellar travel. Following the study of the first known interstellar object to visit our solar system in 2017, named Oumuamua, Israeli-American Harvard astrophysicist Avi Loeb concluded that the object's dynamics could only be explained by the presence of alien technology [169].

10.5 Summary

An advanced alien civilization may decide against sending a signal into space to communicate with other alien species.

Engaging in such information exchange could be perilous for a less advanced civilization. Furthermore, establishing a line of communication between vastly dissimilar alien species may prove exceedingly challenging, if not impossible. An alien species may send unintentional electromagnetic waves as a byproduct of internal communications. However, such signals are probably too faint to be received by an unintended recipient. Furthermore, a sophisticated civilization might colonize new solar systems using artificial intelligence or hitching rides on celestial bodies. The late astronomer Frank Drake even speculated that the number of intelligent civilizations could be equivalent to the number of years spent transmitting signals, perhaps expressing his frustration with estimating the parameters of his renowned equation.

At this stage, estimating L is equivalent to guessing since we have no ground to estimate the average lifespan of an alien civilization. The duration of communication could be 100 years or 10 billion years. Just like in the chapter on intelligence, we have decided to leave this parameter undetermined to consider different cases and the relationship between the duration and the intelligence factor.

$$L = ? \tag{10.3}$$

11

The Drake Equation: Putting All Together

Let us summarize our estimates of the parameters of the Drake equation as follows:

- R is the average annual rate of star formation in the Milky Way: conservatively, we can set it equal to 1.
- f_p denotes the fraction of those main sequence stars in the Milky Way that have planets orbiting them: also conservatively equal to 100%.
- n_e represents the average number of planets that can potentially support life given a solar system with at least one planet: n_e should be about 10%. This number means that some solar systems may not have habitable planets. We select the smaller number to be very conservative, and many other authors have set it higher than unity.
- f_l is the fraction of planets per solar system that will, in principle, develop life among those who can already support it: $\approx 100\%$. This estimate may be more aggressive than others. However, there are clues that this assumption may not be so far-fetched. For instance, life started on Earth almost immediately compared to cosmic times. Furthermore, there is no final evidence that the confirmed exoplanets are suitable for life, hinting that life-supporting planets may be uncommon. The lack of life-supporting exoplanets may support our claim. However, finding a lifeless planet with identical conditions to ours may lower our estimate substantially.
- f_i is the fraction of those planets where evolution creates intelligent life: we have no estimate for this biological parameter.
- f_c is the fraction of advanced extraterrestrial civilizations from which we could receive a signal: $\approx 100\%$. The caveat is that communication is unin-

© The Author(s), under exclusive license to Springer Nature Switzerland AG 2025
L. Vacca, *Life Beyond Earth*,
https://doi.org/10.1007/978-3-031-81695-6_11

tentional, as most civilizations could have no interest in establishing communication with us.

- L is the length communication parameter for which we also do not offer an estimate.

Summarizing, the simplified formula for the number of intelligent civilizations that could communicate with us is equal to

$$N \approx 0.1 * f_i * L \tag{11.1}$$

Generally speaking, N is an estimate for extraterrestrial civilizations, not including us. Suppose the parameter N includes us because we sent signals into space. In this case, N should be equal at least to 1. Now, assuming certainty in the intelligence parameter f_i, the timespan of communication L should be at least equal to ten years, a pretty short interval for an intelligent civilization. Put briefly, we recovered the minimum timespan for galactic communication. This timespan is shorter than how long radio waves have been sent into space. In reality, the intelligence parameter is certainly smaller than unity, but it is not clear by how much. Indeed, we have seen that intelligence has appeared on Earth only quite recently when we compare it to when the earliest life forms appeared. For instance, if we assume that life appeared 24 h ago, early humans discovered fire about 20 s ago. As the probability of intelligence decreases, the communication length lengthens for a fixed N. Hence, when N is fixed and greater than 1, the extremes are (a) the intelligence step is rare, but aliens have been transmitting for long periods, or (b) the intelligence step is common and is coupled with short periods of communications.

What about if N is large? In that case, L should equal or exceed our broadcasting time. From this assumption, one could infer that such civilizations would be, on average, much older and more advanced than us.

11.1 What About Other Exoobjects?

The unknowns in this scenario are the moons tied to giant planets. Moons could be a significant player in the game of life, a hypothesis that still needs confirmation or disproves. The initial Drake equation concentrated exclusively on planets orbiting stars. However, primary life forms or life-building blocks could be present on such celestial bodies. Hence, upon their inclusion in the Drake equation, the value of N may receive a boost.

11.2 The Intelligence Parameter

Above, we set the intelligence parameter to unity to get a sense of the value of N. The intelligence parameter is perhaps the most important of the Drake parameters since a significantly small value would imply that we are probably alone in the galaxy as an intelligent species. While transmission time can continue for the duration of the sending civilization, the intelligence parameter is constrained by the laws of nature and cannot be altered. Conversely, an intelligence parameter approaching or equal to unity would suggest the existence of multiple intelligent species within the Milky Way actively transmitting signals, both intentionally and unintentionally. This raises the classic Fermi question: why don't we hear them? I opine that a simple analogy can shed light on this question as far as communication is concerned. People will probably hear us if we cry for help in a densely populated city. In addition, let us assume that our cry may only be intended for those nearby for two reasons: (A) our cry has limited power and range, and (B) we only require assistance from those nearby. Maybe we need protection from a threat. Individuals located a mile or more away from us will not hear our cry as they are too distant to provide aid.

In contrast, when we call for help, we seek assistance from a particular distant individual or group and no one else. The message reaches its intended recipient regardless of their proximity to us. An example could be calling a doctor or law enforcement.

A possible answer is that Earth is probably in a specific location in the Milky Way, where there are few or no "shouting" civilizations. At the same time, we are not being phoned up by distant advanced civilizations eager to talk to us.

11.3 Wrapping It All Up

By now, it should be clear to the reader that reliable estimates of the parameters of the Drake equation are possible only for the cosmological parameters of the equation.

These parameters benefit from empirical data, and estimates are expected to become notably more precise with the construction of increasingly powerful telescopes. On the other hand, estimating the biological parameters of the Drake equation, especially the intelligence parameter, poses significant challenges, as any educated conjecture could diverge by several orders of magnitude from its actual value. Nonetheless, given that the simplest forms of life emerged on Earth billions of years ago, and considering that the laws of nature are uniform throughout the Universe, it is reasonable to presume that life-generating events occur elsewhere in the cosmos wherever favorable conditions exist. In any case, assuming that $L = 100$, roughly the timespan during which we have been sending radiowaves into space, our estimate for the Drake equation reduces to $N = 10 * f_i$. If $f_i << \frac{1}{10}$, we are probably alone in the galaxy. It is presumable that we are an early young technological civilization and that any advanced civilization is confined to another galaxy.

If we are alone, that would be where extinction events are wiping out all life entirely and rendering a planet inhabitable, as in the case of a supernova event nearby or a direct, powerful gamma-ray burst. Alternatively, civilizations may encounter early demise due to a Great Filter, a highly destructive technology for their creators. But as we have seen in our considerations of non-final extinction events, life demonstrates remarkable resilience, and given sufficient time, it tends to evolve into more intelligent, sophisticated forms. Summarizing, if the intelligence parameter is several orders in magnitudes smaller than unity, we are probably alone in our galaxy:

$$N = 1 \qquad (11.2)$$

Conversely, if the intelligence parameter is within one order of magnitude from unity or better, there are at least ten or more intelligent civilizations in our galaxy:

$$N \geq 10 \qquad (11.3)$$

Given our previous considerations, some of these civilizations should be far more advanced than us. In examining the Drake equation, we've already explored some potential explanations for the Fermi paradox. In the upcoming chapters, we shall explore several alternative theories that address this renowned paradox in detail. As we conclude our discussion on the Drake equation, one crucial observation stands out: discovering even a primitive form of life within or beyond the Solar System would strongly suggest that life in the Universe is likely far more abundant than previously thought. Indeed, where there is smoke, there is also fire. Finding microbial life elsewhere in the solar system besides Earth would support our second base case that N is greater than 10.

12

First Hypothesis: We Are Alone

12.1 Introduction

The famous science fiction writer and futurist Arthur C. Clarke stated: "There are two possibilities: either we are alone in the universe, or we are not. Both possibilities are terrifying".

The first intuitive answer to the Fermi paradox is that humans are the only intelligent civilization in our galaxy and, perhaps, in the entire observable universe. In short, we are alone.

This possibility is supported by the absence of any detection or discovery of extraterrestrial intelligence thus far. Notably, an extensive array of radio telescopes continually scans the universe. Numerous state and private entities are dedicated to the pursuit of extraterrestrial intelligence. The SETI Institute was incorporated as a 501(c)(3) California nonprofit organization in 1984 by Thomas Pierson and Dr. Jill Tarter. Other important SETI figures were Frank Drake and Carl Sagan [170], who, while not the founders, made tremendous contributions to the search for extraterrestrial intelligence.

SETI employs over 100 scientists who diligently analyze information received from space. It is one of many listening organizations coordinating a global network of radio telescopes. For instance, the National Science Foundation's Karl G. Jansky Very Large Array (VLA), located in New Mexico, consists of 27 antennas spread over 23 miles of desert. VLA is a radio astronomy facility used for astrophysical and cosmological research. SETI analyses VLA's radio signals to search for transmissions that only extraterrestrial intelligence could originate [171]. In addition to SETI, many research-funded radio telescopes exist, serving purposes ranging from commercial and industrial to military.

© The Author(s), under exclusive license to Springer Nature Switzerland AG 2025   **91**
L. Vacca, *Life Beyond Earth*,
https://doi.org/10.1007/978-3-031-81695-6_12

And yet, when we compare our considerable effort to the sheer vastness of the universe, we realize that some important signal may have escaped us. This hypothesis cannot ever be ruled out. Our universe is just too vast.

What could be the reasons why we humans could find ourselves alone in the universe? Some theories claim that the Earth has rare properties in the universe that are crucial for enabling and nurturing the emergence and evolution of life. Conversely, others argue that the formation of life itself is exceedingly rare. Let us examine the perspectives of a few prominent authors who support the notion that we are likely alone (Fig. 12.1).

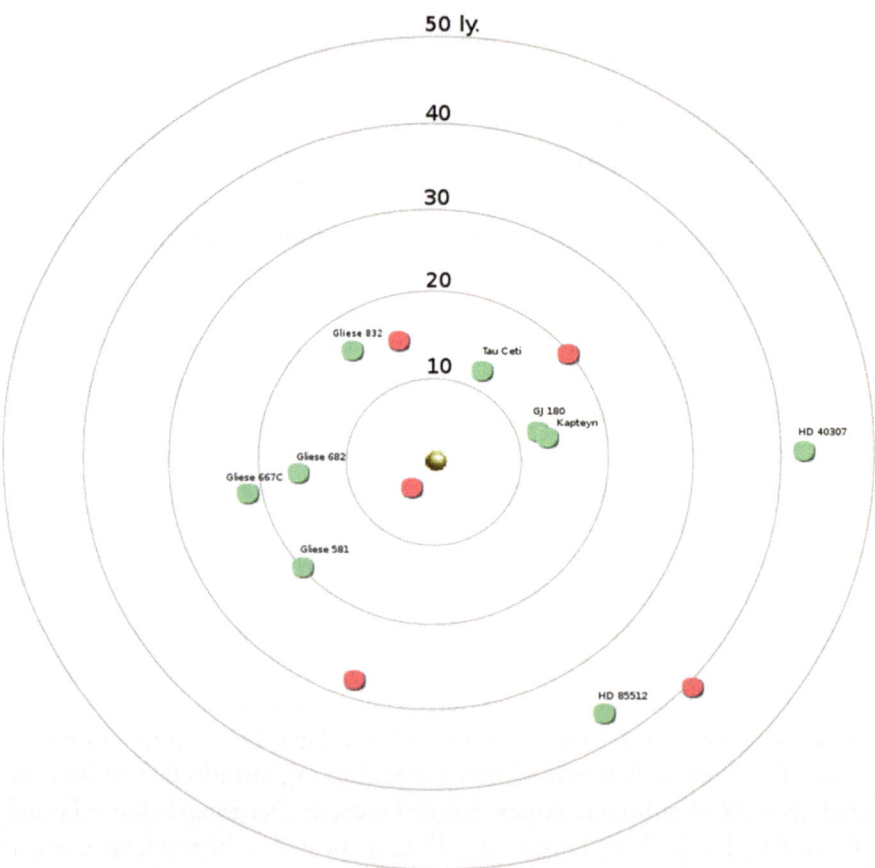

Fig. 12.1 Nearest stars with terrestrial ("rocky") exoplanets candidates at a distance of up to 50 light-years from the Solar System. By Zhitelew, CC0, https://commons. wikimedia.org/w/index.php?curid=33344648

12.2 The Rare Life and Rare Earth Hypotheses

The Rare Earth hypothesis states that Earth is an incredibly unique planet with unique characteristics whose concomitance made life appear about four billion years ago. This hypothesis is frequently conflated or confused with the Rare Life hypothesis, which posits that life, in general, is exceedingly scarce in the universe, even under conditions conducive to its flourishing.

In the past, various influential thinkers have posited that life is rare, if not extremely rare, and its appearance cannot be attributed to our planet. George Gaylord Simpson, a renowned Harvard paleontologist, gained prominence in the early 1960s for critiquing the emerging field of exobiology through a series of lectures. In 1964, he published an essay titled "The Nonprevalence of Humanoids" in the prestigious journal Science [172]. Simpson argued that the emergence of a species like *Homo sapiens* on Earth is more of a highly improbable probabilistic process than a deterministically inevitable event. Hence, according to Simpson, searching for extraterrestrial species like ours on other planets is futile. He contended that the evolutionary steps that led to intelligent life were improbable, from macromolecules such as proteins, lipids, and acids to living cells and intelligent species.

Similarly, German-American biologist Ernst Mayr expressed skepticism about SETI's efforts to communicate with or receive signals from advanced extraterrestrial species [173]. Mayr voiced his views during a public debate with astrobiologist Carl Sagan in 1995.

The idea that exceptionally favorable conditions for life existed on Earth, practically unrepeatable in other parts of the universe, is shared by many astronomers and astrophysicists.

This idea has been covered extensively by American paleontologist Peter Ward and American astronomer Donald Brownlee in their book "Rare Earth: Why Complex Life is Uncommon in the Universe" [174]. Their book makes a case for the Rare Earth Hypothesis. Examples of points made in the book are the right amount of gravity, the presence of a Moon, Earth's quiet location in the Milky Way, plate tectonics, and the existence of gas giants like Jupiter and Saturn in the outer orbits of the solar system, all elements that have favored the evolution process leading to intelligent life. For instance, the position of the gas giants in the outer region of the Solar System has contributed to significantly reducing the number of impacts of asteroids and comets on our planet.

Another book that proposes our planet's exceptionality is "Lucky Planet: Why Earth is Exceptional-What That Means for Life" by David Waltham, a professor of geophysics at the University of London [175]. His thesis is that it will be tough to find exoplanets like Earth because the Earth is a unique planet that is suitable to host complex life. Waltham starts his book by giving an example of a planet that, in the beginning, was like Earth and gradually experienced severe climate change that led to extinction. Such climate has been experienced during major extinctions, but Earth's biosphere has always returned to favorable conditions after millions of years. Waltham's central argument is that Earth is lucky because its climate has been relatively stable through natural and artificial variations. Furthermore, Waltham states that our biased observations deceive us, a form of anthropic selection induced by our senses.

Indeed, one of the main arguments that proponents of the exceptionality of life on Earth is that our condition tends to overestimate the probability of events and causes that are conducive to life. Technical arguments supporting the Rare Earth hypothesis are extensive and could fill many books and articles. Here, we outline a few of these arguments, some of which have already been briefly mentioned in more detail.

12.3   Stability of Gravitational Systems

We begin with the idea that the Earth has been in a stable orbit for billions of years. Simulations show that it will probably continue on a stable orbit for one hundred million years, perhaps for billions of years. It is pertinent to note that the issue of Earth's orbit stability is intricately linked to the stability of the entire solar system. There are primarily two significant sources of uncertainty in this regard. The first source stems from chaos theory and is inherent in all gravitational systems with more than two bodies. In such complex systems, instability may arise over extended periods of many millions of years. This means that the orbits of planets could deviate significantly from their original trajectories due to the introduction of one or more initial small perturbations. According to the findings of French astronomer Jacques Laskar and his colleagues, the dynamics of the solar system are chaotic. Furthermore, the same author has found that it is practically impossible to find precise solutions to the orbital parameters of the inner planets in our system when projected to times exceeding 100 million years [176]. Nevertheless, our solar system has maintained stability for billions of years and is expected to continue for an extended period.

The second source of uncertainty is gravitational disturbances caused by a large planet or star passing near or through our solar system. Such an event could alter the orbits of one or more planets, leading to significant disruption within the entire system. Additionally, the dynamics of the galaxy itself can exert influence over binary or higher multiplicity systems, particularly when the stars are situated at considerable distances relative to our solar system. The field of celestial mechanics presents significant challenges. It revolves around solving systems of differential equations, either using the Newtonian gravity formula established by English mathematician Isaac Newton in 1687 or general relativity, which was developed by Albert Einstein in 1915. Analytical solutions for the full problem of a system with three or more bodies cannot be derived.

Hence, studying the orbital stability of a system such as ours or similar to ours requires the integration of the equation of motion on powerful computers.

Generally, every complex system has a typical timescale, after which the system becomes unpredictable.

This timescale is referred to as a Lyapunov timescale, named after the Russian mathematician Aleksandr Lyapunov, who extensively investigated the stability of dynamical systems [177]. The fundamental question revolves around whether star systems with two or more stars can, under certain conditions, remain stable for a sufficiently long period to allow for the development of life. The answer to this question is affirmative, albeit with a caveat. The caveat is that systems with more stars are more prone to becoming chaotic. Let's consider some systems that have maintained stability for significant durations.

12.3.1 Binary Star Systems

We have seen that binary star systems assume importance because they comprise a large fraction of all star systems in our galaxy; the analysis of a sample of stars states that the fraction is about 48% [178]. Regarding the planets' orbits, there are two kinds: the S-type and the P-type, also known as circumbinary orbit [179].

S-type orbits occur when a planet orbits only one of the two stars in a binary system, while P-type or circumbinary orbits involve a planet orbiting both stars. According to a 2007 study by Quintana and Lissauer, approximately 50–60% of binary systems can potentially host planets with stable orbits [180]. A simulation study shows that circumbinary planets in the external regions of the system must have orbits that are multiples of the distance between the two stars to maintain stability [181]. For example, Kepler-47 is a binary star system located over 3,000 light-years away from Earth, known to harbor three giant planets [182]. Among these planets, Kepler-47c, roughly five times the size of the Earth, possesses a circumbinary orbit in the habitable zone. While a giant planet may not support life, its moons could theoretically provide habitable environments.

12.3.2 Ternary Star Systems

Ternary systems are rarer than binary systems; the study above on star multiplicity finds that about eleven percent of all star systems are ternary systems. Numerical studies involving the integration of planetary orbits in star systems with three stars are very complex and encompass a myriad of configurations. Here, it is essential to note that only hierarchical three-star systems have been observed thus far [183]. Such systems consist of two nested subsystems: one given by an inner binary star system and the other by an outer star. A hierarchical system is stable because the outer star has negligible gravitational interaction with the binary subsystem. For instance, consider the nearest star system to us: *Alpha Centauri*. This is a ternary star system where Alpha Centauri A and Alpha Centauri B have masses comparable to our sun's mass and form a binary system. Conversely, the third star of the Alpha Centauri system, namely *Proxima Centauri*, is a red dwarf star much farther apart from the others. *Proxima Centauri* is the closest star to our solar system at a distance of only 4.2 light years. The age of the Alpha Centauri system, which is slightly older than our solar system, is estimated to be between 4.5 and 5 billion years [184]. Consequently, in principle, it should be possible to identify numerous ternary systems that have remained stable for billions of years.

12.3.3 Quaternary Systems and More

Quaternary star systems represent only a tiny fraction of the total solar systems in the Milky Way galaxy. An example of a quaternary star system that has maintained planetary stability for approximately two billion years is the Kepler-64 system [185]. Situated approximately 5000 light-years from Earth according to the Open Exoplanet Catalog, this system is known to harbor at least one planet: Kepler-64b, a gas giant orbiting two of the stars within the Kepler-64 system. The discovery of Kepler-64b was made by two volunteers, the so-called Planet Hunters, a citizen science project. The discovery was announced in 2012. Systems featuring five or more stars are considerably rarer in comparison.

12.3.4 The Problem of Orbital Stability and Life

The problem of orbital stability in a star system is highly complicated. All complex systems are susceptible to chaotic dynamics over extended periods. While examples of solar systems with multiple stars that have remained stable long enough for life to emerge exist, it's important to note that systems with only one star are generally more stable than those with multiple stars.

12.4 The Presence of an Atmosphere with Oxygen

Earth, Venus, Mars, and the gas giants in our solar system all possess an atmosphere [186]. Several factors impact the presence of an atmosphere on a celestial body: the mass of the body, the chemical makeup of the atmosphere, its temperature, its orbital position, and the type of star it orbits. Typically, the greater the mass of a planet, the stronger the gravitational force on its atmosphere. To illustrate, gas giants primarily consist of hydrogen and some helium in their atmospheres. In any case, it is improbable that life will exist on these types of planets due to their lack of surface and chemical composition. Inner orbit planets like Venus and Mars mostly have carbon dioxide and a little nitrogen in their atmospheres [187]. Scientists can determine the chemical composition of exoplanets through spectroscopic analysis of light passing through the atmosphere during transit. Even moons can possess atmospheres. Saturn's largest moon, Titan, boasts an atmosphere denser than Earth's, primarily composed of nitrogen and a little methane. With the advent of the James Webb Space Telescope, launched in December 2021, researchers anticipate more detailed studies of exoplanet atmospheres.

An exemplary system hosting rocky planets akin to Earth in size and atmospheric composition is the Trappist-1 system, located approximately 40 light-years away [188]. Three of these planets reside within habitable zones. The star is an ultracool red dwarf star. Despite the star's cool temperature, the rocky planets orbit close enough to support liquid water. Among those in the habitable zone, the planet Trappist-1e stands out due to its mass and radius similar to Earth's. Its rocky surface suggests the presence of iron, and notably, it lacks a hydrogen-dominated atmosphere, though the presence of any atmosphere on Trappist-1e remains uncertain. Moreover, its proximity to an active star raises concerns about potential atmospheric loss, a challenge prevalent in systems with smaller active stars.

Finally, oxygen in a planet's atmosphere does not always indicate biological activity. A research article in the journal Science Advances by Måns Wallner et al. shows that sulfur dioxide could be an abiotic source of oxygen pumped into the atmosphere through volcanic activity [190]. Therefore, it is entirely possible that typical biosignature elements and compounds can have an abiotic source, i.e., not produced by life.

12.4.1 Searching for Biosignatures

The search for an Earth twin is still ongoing. While many rocky Earth-like exoplanets were discovered in habitable zones, to the author's knowledge, none of these exoplanets have atmospheres, pressures, and temperatures similar to our planet's atmosphere. One aspect that makes this search harder is that the effective transit search method tends to find much more giant planets than Earth and much closer to their stars. Nevertheless, it is important not to dismiss the likelihood of atmospheres existing in certain sizable moons and the potential for life forms that don't require high oxygen levels in their atmospheres to survive. These life forms might adapt to environments vastly different from ours.

12.5 The Presence of a Protective Magnetic Field

Electric currents within Earth's iron core produce the Earth's magnetic field. This field extends into space and functions as a shield to the Earth from solar-ionized particles by deflecting them. This deflection mechanism operates based on the physics principle that charged particles alter their initial trajectories as they tend to follow the lines of a magnetic field. It's widely understood that ionizing radiation threatens life as it can change or break the molecules of living cells. Additionally, Earth's magnetic field serves to protect our atmosphere by reducing the erosion caused by solar-charged particles.

What are the chances that a rocky exoplanet could possess enough iron to generate its substantial magnetic field? Firstly, all the rocky planets in our solar system, including the moon, have iron cores. However, what about exoplanets? The iron content of an exoplanet or exomoon depends on its host star's metallicity. Metallicity refers to the abundance of elements heavier than hydrogen and helium in a star. It is often quantified as the ratio between a star's iron and hydrogen content. Studies have demonstrated that higher iron content in a star corresponds to a more significant presence of iron in its rocky planets [191]. Moreover, star metallicity correlates directly with the likelihood of having gas giants in the star system [192]. Given that our Sun exhibits high metallicity, it harbors gas giants and rocky planets with iron cores. Research into magnetic fields on rocky planets is still in its early stages. Notably, astronomers Pineda and Villadsen have made significant contributions, detecting the first signs of a magnetic field emanating from a planet known as YZ Ceti b [193]. This rocky planet orbits the star YZ Ceti, 12 light years away from Earth. Their observations of a radio signal from YZ Ceti suggest a possible magnetic interaction between the star and its planet.

While the presence of a rocky exoplanet with a magnetic field needs further confirmation, methods such as the one employed by the astronomers above could be used to investigate other exoplanets.

12.6   The Solar System in the Milky Way Galaxy

The Milky Way, a spiral galaxy, houses our solar system within one of its inner arms, the Orion arm. Our solar system is roughly 26 thousand light years from the galactic center [194]. Considering the galaxy's diameter spans approximately 100 thousand light years, our solar system is somewhat peripheral. This positioning may have contributed to the emergence and sustenance of life on Earth. At the heart of the Milky Way lies Sagittarius A*, a supermassive black hole boasting a mass equivalent to about four million suns [195]. The central region of our galaxy is extremely densely populated with stars, much denser than the region around the Sun, with millions of stars in a squared volume of the size of a parsec [196]. Consequently, this area is characterized by intense star formation, jets of hot gasses, and elevated levels of harmful radiation, including ultraviolet, X-ray, and gamma rays, which pose significant threats to life. Moreover, the central stars, often massive, can lead to devastation upon their supernova explosions at the end of their life cycles.

However, the location alone is not the sole determinant; the activity of stars plays a pivotal role. For instance, our supermassive black hole at the center of the Milky Way is relatively dormant, but it has a recent past of eating objects passing by. This combination of a peripheral position and subdued stellar activity within our galaxy has likely fostered the emergence and evolution of life, nurturing it toward its current complexity on Earth. Nonetheless, the rarity of such favorable conditions in other galaxies remains an open question. It is conceivable that similar conducive environments for life may exist in regions of the cosmos where galaxies are less densely populated.

12.7   The Moon and Its Impact on Life

Contrary to popular belief, seasonal changes do not arise from variations in the distance between the Sun and Earth throughout the year. Earth's orbit, albeit slightly elliptical, exhibits a minor degree of eccentricity. It is almost circular. Seasons, instead, are caused by the change in the tilt of the Earth's axis.

The Earth's axial tilt causes seasonal variations, with summer occurring when a hemisphere tilts toward the Sun and winter when it points away. As proposed in 1946 by Canadian-American geologist Reginald Daly, approximately 4.5 billion years ago, the Moon formed from a collision between Earth and a Mars-sized planet. This theory is called the giant-impact hypothesis [197]. The impact resulted in the ejection of molten material into space. This material coalesced into the Moon, which now orbits Earth at approximately 400 thousand kilometers. Gravitational forces have locked the Moon's rotation, causing it to always present the same face to Earth. Due to this collision, Earth experiences a slight axial tilt, akin to the wobble of a spinning top, which is kept in check by the Moon's gravitational influence. The Moon's gravitational pull also creates ocean tides, crucial for marine life in intertidal zones. Another potential benefit of having a moon is its influence on Earth's magnetic field. Research from CNRS and Université Blaise Pascal suggests that the Moon's tidal forces strengthen Earth's magnetic field, which is crucial for shielding life from harmful cosmic radiation [198]. The prevalence of moons on rocky planets is relatively uncommon. Mars has two moons, Phobos and Deimos, while Mercury and Venus lack moons altogether. Gas giants typically have numerous moons due to their substantial mass, making them more likely candidates for moon presence. Thus, Earth-sized rocky exoplanets have a lower likelihood of possessing moons. Research by Professor Miki Nakajima at the University of Rochester indicates that terrestrial and icy exoplanets with a radius not larger than 1.6 times the radius of the Earth are likely to acquire fractionally larger moons during interplanetary collisions [199]. That would explain the fact that we have such a large moon compared to the size of the Earth. Therefore, searching for potential life should prioritize rocky exoplanets with masses similar to or slightly greater than Earth's, as they will likely form fractionally large moons.

Additionally, it is essential to point out that collisions between celestial bodies are expected during the formation of solar systems, which heightens the likelihood of rocky planets acquiring moons. Among moons in the Solar System, our Moon stands out as the most massive relative to its parent planet. The size, mass, and proximity of the Moon to Earth are pivotal factors in its influence on the existence of life on our planet. Further research is necessary to determine the uniqueness of these characteristics for rocky exoplanets within our galaxy and beyond.

12.8 Gas Giants on the Outer Region of the Solar System

Gas giants like Jupiter and Saturn have played and continue to play an important role in protecting life on Earth from asteroid and comet collisions. They deflect and eject such objects thanks to their powerful gravitational attraction.

The impact of asteroid collisions has been significant throughout Earth's history, causing both major and minor mass extinctions. These asteroids, remnants from the solar system's formation 4.5 billion years ago, are primarily found in the asteroid belt between Mars and Jupiter. Ranging in size from small rocks to massive objects like the Chicxulub asteroid, which was ten kilometers in diameter, these asteroids pose a considerable threat to Earth. NASA reports that a significant portion of meteoric collisions originate from this asteroid belt. On the other hand, most of the comets originate from two principal regions: the Oort cloud and the Kuiper belt. Comets are small, icy bodies that develop long tails as they approach the Sun. The Oort cloud, an expansive region extending up to 100,000 AU from the solar system and believed to be the source of comets with long orbital periods, is influenced by passing stars or other massive objects [200]. The Kuiper belt, located closer to the Sun than the Oort cloud and extending well beyond Neptune, is another source of comets with shorter orbital periods. How rare would it be to locate gas giants in the outer regions of other solar systems? Regarding gas giants in other solar systems, current observations suggest they are typically found close to their stars. However, this distribution may be influenced by observational biases inherent in current exoplanet detection methods. This finding supports the idea that the configuration of our solar system is rare and constitutes a strong argument in favor of the rare earth hypothesis. One plausible hypothesis suggests that gas giants initially formed as rocky, icy planetary cores near their young stars. Over time, these cores accumulated significant amounts of hydrogen and helium from the surrounding protoplanetary disk, ultimately transforming into gas giants. Another potential scenario proposes that certain gas giants originated in the outer regions of their solar systems and subsequently migrated inward toward their stars, gradually accreting additional material along the way. Notably, astronomers have found a relatively small number of gas giant exoplanets very far from their sun. For example, researchers Markus Janson and collaborators published the existence of a wide-orbit giant planet in the high-mass b Centauri binary system in 2021 [201]. Finally, given the limited number of exoplanets discovered so far, new methods of exoplanet search may enable us to find more gas giants on the outer skirts of their solar systems.

12.9 Plate Tectonics

Plate tectonics is a theory about forming continents, mountain ranges, volcanoes, and earthquakes. It was formally introduced in 1912 when German geologist Alfred Wegener proposed that continents are plates that gradually drifted apart from one giant supercontinent known as Pangaea [202]. In a nutshell, Earth's crust comprises a series of plates that are continuously moving and forming the continents. The lithosphere, i.e., the crust and the uppermost rocky part of the Earth's mantle, comprise the vast continental plates that have been moving for 4 billion years. The plates are vast, rigid slabs that move from one to several centimeters annually, one relative to the other, and rest on the asthenosphere, a layer of molten rock [203].

Tectonic activity is crucial in sustaining Earth's life cycles, driving processes like the carbon, nitrogen, and water cycles. For example, carbon, primarily in the form of carbon dioxide (CO_2), is released into the atmosphere through volcanic activity. Plants and algae then absorb carbon dioxide during photosynthesis. Additionally, ocean water absorbs carbon dioxide from the air, which can form carbonic acid (H_2CO_3). Marine organisms create shells and skeletons by combining calcium and carbonate ions to create calcium carbonate ($CaCO_3$), insoluble in water [204]. Lately, this exchange of CO_2 between the atmosphere and the oceans has been disrupted due to an excess production of CO_2 due to burning fossil fuels.

Plate tectonics continuously draws the elements of the Earth's crust back into the mantle in a process known as *subduction*. Earth's mantle is a solid layer that constitutes most of our world. Among the elements of the mantle, we find carbon. A lot of carbon dioxide is necessary to trap heat in the atmosphere, but an excess leads to the greenhouse effect and contributes to global warming.

As well known, Earth's atmosphere comprises about 78% nitrogen, 21% oxygen, 0.9% argon, 0.03% carbon dioxide, and trace amounts of other gases.

Nitrogen was released into the atmosphere during past volcanic activity. Plate tectonic dynamics play a pivotal role in the deep water cycle, as water is drawn down into the mantle through the subduction of tectonic plates and eventually returned to the oceans. Rocky planets like Venus and Mars currently lack tectonic plates. However, there is evidence that Venus had tectonic plates in the past [205]. The basis of this finding is the significant presence of molecular nitrogen and carbon dioxide in the atmosphere and the surface pressure. The authors claim that a single-plate stagnant lid regime cannot explain such elements.

Evidence of a plate tectonics system has been found in Europa, one of Jupiter's four massive Galilean moons. American geologists Simon Katterhorn and Louise Prockter discovered signs of subduction on Europa's icy shell, as detailed in their article in Nature [206]. Europa harbors a vast salty ocean beneath its surface, and exomoons orbiting gas giants experience tidal heating due to the gravitational forces exerted by their parent planets. Similarly, an exoplanet close to its host star may undergo tidal stress, potentially driving plate tectonics. For instance, researchers from the University of Bern, led by Tobias Meier, identified evidence of tectonic activity on the distant exoplanet LHS 3844b, orbiting the star LHS 3844, located 48.5 light-years from Earth [207]. However, due to its extreme temperatures and lack of atmosphere, LHS 3844b is uninhabitable.

At this time, very little is known about the frequency of plate tectonics on rocky exoplanets and exomoons.

12.10   Summarizing

Undoubtedly, given our limited sample of discovered exoplanets, Earth exhibits a combination of characteristics that appear exceptionally rare. Whether these traits will remain rare as we discover more exoplanets is uncertain. Individually, isolated characteristics are already observed on one or more exoplanets. However, exoplanets potentially possessing one or more of these traits often prove uninhabitable for various reasons. They may be excessively hot because they are positioned too close to their host stars or too cold if far away. They may lack an atmosphere or be vulnerable to solar flares. At the very least, an exoplanet capable of hosting life should be a rocky planet situated within a habitable zone, orbiting a stable sunlike star. It should possess an iron core and an atmosphere containing significant amounts of oxygen, carbon dioxide, and ample liquid water.

The Rare Earth hypothesis remains a plausible explanation, suggesting that life is indeed rare within our galaxy due to the scarcity of Earth-like planets with solar systems similar to ours. If we are alone in our galaxy, it is probably because life is rare. The vastity of our galaxy makes it improbable that there are no Earths out there.

It is worth noting that life on other celestial bodies may differ significantly from terrestrial life and may not rely on carbon as its fundamental building block. Some biochemists have proposed silicon as a potential alternative to carbon in forming life.

In conclusion, the Rare Life and Rare Earth hypotheses suggest a universe where life is exceedingly rare, making communication highly improbable.

12.10 Summarizing

13

The Great Filter Hypothesis

American economist Robin Hanson introduced the "Great Filter" concept in an article titled "The Great Filter-Are We Almost Past It?" published in 1996 [209].

The concept of the Great Filter suggests that extraterrestrial civilizations may never reach a stage where they can colonize their galaxy and be detectable because they encounter obstacles or setbacks that impede their technological progress or lead to their early demise. This idea posits a fundamental challenge that all intelligent life forms will likely face at some point in their development. The existence of such a filter could explain why we have not yet found any signs of extraterrestrial civilizations.

What is the nature of the Great Filter? Is the Great Filter behind us, or will humanity have to confront it in the future?

Expanding this concept, we can apply it to intelligent life and all life forms, ranging from microorganisms to advanced civilizations like Type III on the Kardashev scale.

First, what factors might prevent or have prevented the emergence of life on a planet?

13.1 Looking for Great Filters in the Past

The Rare Earth Hypothesis constitutes a barrier to intelligent life because conditions on an exoplanet may change in a manner that endangers all kinds of life. Under this assumption, life cannot develop further because the planet or moon loses the conditions for life to evolve and flourish. An example could be a planet expelled from its habitable zone due to a gravitational interaction.

© The Author(s), under exclusive license to Springer Nature Switzerland AG 2025
L. Vacca, *Life Beyond Earth*,
https://doi.org/10.1007/978-3-031-81695-6_13

The Rare Life Hypothesis presents another significant filter, which lies in our past. Suppose abiogenesis was the only mechanism for the emergence of life on Earth 3.5 billion years ago. In that case, the Rare Life Hypothesis becomes less plausible because the origin of life is intricately linked to our planet and its unique conditions. It is entirely feasible that life could emerge frequently on planets or moons similar to ours. Conversely, suppose panspermia played a role in seeding life on Earth, as some meteorite evidence suggests. We might expect panspermia to occur widely throughout the universe, not just in our local vicinity. Consequently, we would anticipate that the seeds of life could have been dispersed to many exoplanets through panspermia, potentially leading to a proliferation of life in our galaxy, a phenomenon that we have yet to observe. In this scenario, the Rare Life Hypothesis becomes more plausible since panspermia has not yielded the expected proliferation of life.

The evolutionary journey of life on Earth, from simple protocells to humans, can be described as a series of incremental steps. These steps could have denoted a potential Great Filter event on another planet or moon.

13.1.1   What Is Life?

One of the founding quantum mechanics scientists, German physicist Erwin Schrodinger, penned a seminal work in 1944 titled "What Is Life? The Physical Aspect of the Living Cell" [210]. This book contains his disquisitions into the interplay between physical principles and biological phenomena. He begins by noting nature's ability to create order out of chaos. He cites examples such as the deterministic heat conduction equation in statistical mechanics, which explains heat transfer through the random movement of countless atoms. However, Schrödinger highlights a crucial distinction: while disorder can lead to order in certain physical systems, life requires a different kind of order–a structured arrangement within a limited number of molecules. This insight leads him to propose that the order observed in living organisms emerges not from disorder but from pre-existing different forms of order.

His central thesis is that "hereditary material" can only be represented by a stable substance, an "aperiodic crystal." This example is the type of order in which living things can only come from order. Life needs hereditary material, which, in turn, contains the information necessary to copy itself, develop, and metabolize. Schrodinger's concept laid the groundwork for the groundbreaking discovery made in 1953 by American biologist James D. Watson and English physicist Francis H.C. Crick: the double helix structure of DNA, the molecule housing genes [211].

An important consideration Schrodinger raises is how life deals with the second law of thermodynamics. The second law states that entropy, a "disorder" measure, always increases in a closed system when energy is transformed or transferred.

The second law of thermodynamics may challenge the existence of ordered systems such as life. However, Schrodinger argues that there is no conflict. Organisms utilize "negentropy," or negative entropy, as he terms it while releasing heat to the environment. This process ensures that order within the organism does not violate the second law of thermodynamics. Moreover, the storage of genetic sequences can be likened to storing information on a computer. Just as certain information is crucial for the proper functioning of a computer's operating system, genetic information is vital for the functioning of living organisms. Like a computer, there must be a boundary or interface between the cell and its environment. This boundary is maintained by a process known as "homeostasis," which ensures the stability of a cell or a more complex organism [212]. In a cell, homeostasis is facilitated by the cell membrane, which selectively allows nutrients and waste to pass through the barrier. This perspective introduces another potential Great Filter: the emergence of the RNA world and the first protocells.

13.1.2   The RNA World and Protocells

Ribonucleic acid (commonly known as RNA) is a single-stranded nucleic acid that led to its close relative DNA (deoxyribonucleic acid) as the main genetic information replicator.

DNA, a stable nucleic acid structured as a double helix, carries genetic information in all living organisms. Despite its prevalence, RNA remains vital and has not been entirely supplanted by DNA. One of its primary functions is the creation of proteins essential for the growth and maintenance of plant and animal cells [213]. According to the RNA world hypothesis, life on Earth originated from a molecule capable of self-replication through a process known as "autocatalysis" [214]. In biology, the concept of a "protocell" refers to a prebiotic structure or very primitive cell where the replication of genetic material occurs independently [215].

A biological model that may describe a protocell is the *chemoton* model conceived by Hungarian biologist Tibor Gánti in 1952 [216]. The chemoton has three fundamental characteristics: a membrane of lipids (a fat substance insoluble in water), a metabolic system that converts food into energy by expelling waste, and a self-reproducing mechanism. The consensus amongst scientists is that the self-reproducing mechanism in protocells was based on RNA or a similar hereditary mechanism.

How did the RNA world form? The answer to this question is largely unknown. Some scientists think that the prebiotic soup contained all the necessary materials for RNA formation, and these components interacted to produce various forms of replicating macromolecules through a trial-and-error process until a more stable macromolecule emerged. The emergence of the RNA world is a potential candidate for the list of Great Filters. However, since it arose approximately four billion years ago under favorable conditions and relatively quickly compared to the age of Earth, it may be considered a less probable Great Filter.

13.1.3 Prokaryotes

Protocells eventually evolved into prokaryotic cells lacking a nucleus and membrane-bound organelles [217]. Prokaryotes are unicellular organisms of two types: Bacteria and Archaea. Bacteria are minuscule organisms, often invisible to the naked eye, found in soil and water, and present in all animals. Archaea are single-celled organisms that thrive in extreme environments, such as high-temperature locations like Yellowstone hot springs, icy environments like glaciers and polar caps, and highly saline environments like the Dead Sea. They also exhibit high resistance to radiation.

Due to their resilience, these organisms may represent the earliest ancestors of life, if any, on other planets. Archaea is a prime example of how life can flourish and adapt even under the harshest physical and chemical conditions.

Prokaryotes are believed to have appeared on Earth about 3.5 billion years ago [218]. Similar to the RNA world, prokaryotes emerged relatively quickly following the emergence of RNA-based life. Given their rapid development and ability to thrive in diverse environments, it is unlikely that the evolution of prokaryotes represents a significant Great Filter event in the progression of life. Therefore, we conclude that the evolution of prokaryotes is a less probable candidate for a Great Filter.

13.1.4 Eukaryotes

Eukaryotic cells, unlike prokaryotic cells, are characterized by their much larger size and well-defined membrane-bound nucleus. They are the fundamental building blocks of unicellular and multicellular organisms, including plants, animals, and humans. Regarding biomass, eukaryotic cells constitute Earth's most prevalent form of life through plants [219].

The leading theory on how eukaryotic cells formed from prokaryotic cells is known as the *endosymbiotic theory*. An eukaryotic cell emerged when an ancient Archaea cell engulfed a bacterium, resulting in a mutually beneficial relationship [220]. This symbiosis occurred because the Archaea cell's tough membrane protected the bacterium, which eventually evolved into the mitochondrion–a membrane-bound organelle providing energy to the newly formed eukaryotic cell [221]. The first eukaryotic cells appeared between 1 and 1.5 billion years of prokaryotic evolution. They were the first to utilize sexual reproduction, which generates more genetic diversity in offspring compared to asexual reproduction [222, 223]. The transition period from prokaryotes to eukaryotes took at least a billion years, significantly longer than earlier evolutionary steps, such as the RNA world and the emergence of prokaryotes. This prolonged duration likely reflects the increased complexity within eukaryotic cells. Therefore, the emergence of eukaryotes represents one of the most challenging evolutionary milestones of life on Earth, potentially serving as a significant Great Filter.

13.1.5 Multicellular Organisms

According to research conducted by biologist J.T. Bonner on the evolution of multicellularity, multicellular life evolved from unicellular cells at least 25 times beginning about 3 billion years ago [224]. The formation of plants and animals came from eukaryotic cells started a billion years ago [225]; such multicellular organisms formed through cell division or aggregation of many cells. Additionally, the enormous number of unicellular organisms is crucial in the chance encounters required to develop multicellular life. Considering these mechanisms and the various ways multicellular life could arise and the times it evolved, it seems improbable that transitioning from unicellular to multicellular organisms would serve as a significant Great Filter in the evolutionary process (Fig. 13.1).

13.1.6 From Animals to Human Intelligence

It took 700 million years from the formation of sponges [226], purportedly the first type of animal on Earth, to the arrival of the *Homo erectus* and his use of fire. *Homo erectus* dates back to about 2 million years ago, which is equivalent to saying yesterday when compared with the evolutionary stages of life on Earth [227]. *Homo erectus*, utilizing fire for warmth, cooking, and protec-

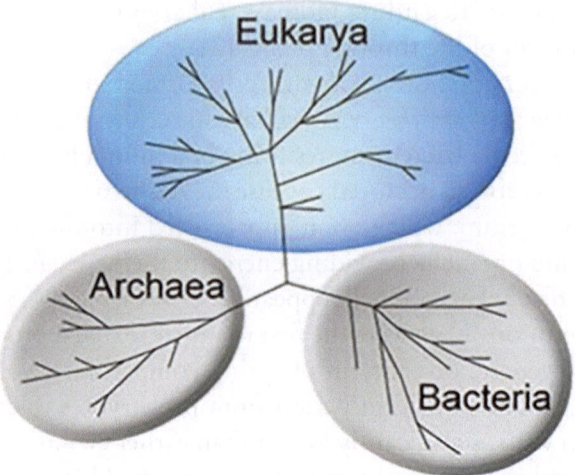

Fig. 13.1 A view of the phylogenetic tree based on the three-domain system, showing the divergence of modern species from their common ancestor in the center. By Michael Stat, Megan J. Huggett, Rachele Bernasconi, Joseph D. DiBattista, Tina E. Berry, Stephen J. Newman, Euan S. Harvey & Michael Bunce—Extracted from this Commons file, CC BY-SA 4.0, https://commons.wikimedia.org/w/index.php?curid=90616393

tion, dispersed across different continents, displaying remarkable adaptation to diverse climates. Additionally, *Homo erectus* exhibited encephalization, signifying an increase in brain size relative to body mass. Supporting a larger body and brain demanded substantial energy. This species demonstrated mobility, seeking out more favorable ecosystems regarding food availability while engaging in hunting activities. From a duration point of view, the transition from simple organisms to intelligent life appears to be an improbable evolutionary step. Another argument supporting intelligent life as a significant filter is the genetic similarity between humans and great apes, with over 90% of DNA shared. However, while great apes continue to inhabit jungles and forests, humans have developed the capacity to build rockets and explore the Moon and beyond. Conversely, humans also exhibit shared behaviors with other social animals. For example, gorillas organize themselves into hierarchical social structures like ours [228], while humans, like hypersexual bonobos found in the Congo, can mate at any time without a specific breeding season.

So, what is the magic ingredient that has made the difference all along? After all, we share many organs with many animals, not just great apes.

Some researchers argue that the ability to walk with an upright posture and manipulate objects with their hands to build and use tools is the defining characteristic that sets our ape ancestors apart from other primates [229].

In Stanley Kubrick's iconic film "2001: A Space Odyssey," a memorable scene depicts an ape, presumably our progenitor, learning to wield a bone as a tool, symbolizing the dawn of intelligence. However, our capacity for tool-making comes with its challenges. It requires larger brains, which consume a significant portion of our energy intake, to function. Additionally, it leads to longer gestation periods and extended periods of infant dependency on maternal care for survival. In 1990, a prominent American evolutionary biologist, Stephen Jay Gould, argued that human intelligence was a stroke of luck [230]. His reasoning stemmed from a hypothetical scenario he called replaying the "life's tape," suggesting that different outcomes could have emerged, resulting in radically different evolutionary paths instead of leading to the emergence of *Homo sapiens*. One possible reason is that stochastic factors influence the evolutionary path in a significant way.

A class of algorithms, the so-called genetic algorithms, are designed to simulate an evolutionary process requiring an objective or goal to be determined. In these algorithms, the goal is called the fitness function, defined at the onset. An example of fitness can be maximizing the probability of survival or the likelihood of reproduction. Notably, such simulations can yield a range of suboptimal points widely spaced apart depending on the fitness function, suggesting that evolution could result in vastly different intelligent species from us. While it is conceivable that local optima may converge to the exact global optimum given enough time, this process could extend far beyond the habitable lifespan of an exoplanet or exomoon. Considering that the Sun is predicted to engulf the Earth in approximately five billion years when it depletes its hydrogen fuel, similar constraints apply to stars more massive than the Sun. Consequently, there is a finite timeframe for the development of intelligence. In summary, the emergence of intelligence represents a plausible Great Filter, potentially limiting the prevalence of advanced extraterrestrial civilizations in our galaxy and beyond.

13.2 Great Filters in the Future

It is not difficult to envision potential Great Filters in our future. One such filter could be global warming, which poses a significant threat to any sufficiently industrialized extraterrestrial society. The release of gases and substances into the atmosphere, land, and water could lead to unbearable levels of planetary heating. According to the laws of thermodynamics, the generation of electric energy inevitably results in residual heat that must be dispersed into the environment. Therefore, addressing global warming is crucial for humanity's continued existence. Another potential future Great Filter is the prospect of a global nuclear war. With superpowers stockpiling a vast number of thermonuclear bombs, the detonation of these weapons on a large scale could trigger a "nuclear winter." Similar to the volcanic winter mentioned earlier, this scenario could result in widespread cooling and heightened radiation levels across the globe. Furthermore, several other technologies, including nanotechnology, bioengineering, biological weapons, and artificial intelligence (AI), have been identified as potential future Great Filters. The misuse or mishandling of these technologies could lead to catastrophic consequences, emphasizing the need for responsible and ethical development and implementation.

AI, in particular, has been the object of many speculations that it may become sentient and replace us.

Anyone who has watched movies like Terminator or The Matrix has witnessed the horrific portrayals of wars between humanity and machines. With today's advancements in robotics and AI, the apocalyptic futures depicted in science fiction movies could become a reality. Another potential Great Filter is a future technology or experiment whose sole existence could endanger life on a planet, as proposed by Swedish philosopher and futurist Nick Bostrom [231]. Some years ago, it was speculated that experiments conducted at the Large Hadron Collider could generate a black hole on Earth, leading to catastrophic consequences. While theoretically possible according to Einstein's equivalence of mass and energy, scientists have emphasized that Earth is constantly bombarded by particles traveling at speeds close to that of light. Yet, as far as we know, no such collisions have produced black holes. Moreover, cosmic events from the past pose ongoing threats that could materialize for every exoplanet, including our own. These threats include supernovae, gamma-ray bursts, asteroid impacts, collisions between moons and planets, solar flares, the approach of stars, neutron stars, black holes, and magnetars, as well as chaotic instabilities within solar systems, to name a few. To illustrate, magnetars, neutron stars with the strongest magnetic fields in the universe, pose a significant threat to life if they come too close to an inhabited star system. Past large extinction events on Earth have been attributed to supervolcanic eruptions.

If supervolcanoes like the Yellowstone Caldera in Wyoming or the Phlegrean Fields in Italy were to erupt at full force, they could trigger a volcanic winter across the planet. However, since life on Earth has endured such events, we consider them less likely Great Filters. Considering the scale and severity of potentially catastrophic events that any extraterrestrial intelligent life might encounter on their exoplanet, their survival ultimately hinges on their ability to colonize new worlds and expand throughout their galaxy. Throughout billions of years, when the probabilities of cosmic threats become significant, space exploration and colonization become imperative for a highly advanced society. In contrast, the inability to undertake widespread space travel at tremendous speeds to reach new habitable worlds could serve as a formidable Great Filter. The absence of encounters with extraterrestrial spaceships (often referred to as UFOs in sci-fi lore) in space could suggest the immense difficulty, if not impossibility, of interstellar travel. As a case in point, a civilization seeking new habitable locations from nearby stars might opt to settle on Mars and transform it into their future home through terraforming efforts. However, we have not found life on Mars, let alone an intelligent one.

Swedish philosopher Nick Bostrom has proposed a theory suggesting that the absence of evidence of life on Mars and in the Solar System implies that the Great Filter lies behind us [232]. Hence, the lack of evidence for extraterrestrial life in our solar system and elsewhere should be considered good news. According to Bostrom, discovering evidence of past life on Mars would suggest that life is more widespread in the galaxy than previously believed. As they say, where there is smoke, there is fire. However, since we have not encountered any extraterrestrial civilizations, Bostrom suggests that intelligent life may have faced a Great Filter that led to its demise, and we are yet to confront it.

Bostrom's argument is plausible but not entirely convincing, in my opinion. Our galaxy is certainly incredibly vast, and much needs to be discovered. Not finding life in our Solar System is no proof that there is no life out there. SETI's astrobiologist, Jill Tarter, and her collaborators once wrote that the SETI search is so minuscule that it can be compared to "having searched a drinking glass's worth of seawater for evidence of fish in all of Earth's oceans" [233].

14

The Zoo Hypothesis

14.1 The Problem of Extraterrestrial Contact

When asked by a journalist whether we should attempt to communicate with extraterrestrial civilizations, Hawking expressed a caution that cannot be overstated. He warned that establishing contact could prompt an advanced alien species to visit Earth, a scenario that could have catastrophic outcomes. His comparison to Columbus's landing in America, which didn't turn out well for the Native Americans, serves as a stark reminder of the potential dangers. Hawking's warning about the consequences of contact between vastly different alien civilizations is not unfounded. The potential for catastrophic outcomes, particularly for the less technologically advanced civilization, is a stark reality. In such a scenario, the fear of destruction or subjugation is a sobering reminder of the risks involved in inter-civilizational contact. It's crucial to approach such potential encounters with caution and a deep understanding of the possible consequences.

Hawking's concerns find striking parallels in human history, where events such as the displacement and conflicts experienced by Native Americans following the arrival of European settlers, beginning with Christopher Columbus's voyage in 1492, serve as poignant and relatable examples. These historical events, while tragic, are crucial in helping us understand the potential consequences of contact between highly different civilizations. Only by learning from these past experiences can we be better prepared for possible future encounters. Another significant historical example of a technologically superior civilization encountering a significantly less advanced one with devastating results occurred when the Spanish conquistador Francisco Pizarro arrived in

L. Vacca, *Life Beyond Earth*,
https://doi.org/10.1007/978-3-031-81695-6_14

Peru. At that time, the Inca empire governed a vast territory encompassing modern-day South America, including Ecuador, Peru, and Chile.

With the assistance of a few hundred armed men along with allied belligerent tribes, Pizarro seized the Inca emperor Atahualpa in the Battle or Massacre of Cajamarca in 1532 [234]. Within a century, Spanish royal authority had gained complete control over Inca territory, leading to the resettlement of indigenous populations. In both cases, native communities were also exposed to new diseases introduced by European invaders. Lacking immunity to these illnesses, the native populations suffered greatly from disease outbreaks. A similar story occurred with the Indian wars. In the nineteenth century, Indian populations suffered tremendous loss of lives due to battles with settlers and the US armies, as well as from disease. However, returning to the prospect of a potential alien encounter, such an event could offer benefits where there is a peaceful exchange of scientific and technological knowledge. Drawing from historical examples, an instance of collaboration between civilizations is seen in the Silk Road, a network of routes connecting China to the Mediterranean, used for trade and commerce from the Han dynasty in 130 B.C. until around the 1400s with the rise of the Ottoman Empire. This exchange involved material goods like Chinese silk, tea, and spices imported to the West and facilitated cultural and religious exchanges.

We have discussed how a technological gap, differing biological environments, and the introduction of foreign biological agents can create imbalances between alien civilizations. In some instances, the meeting of civilizations results in a scientific and technological exchange, leading to a shift in the cultural landscape for both parties involved. However, there are cases where a civilization may opt to preserve its cultural identity and way of life by avoiding external exchanges, even if such interactions could offer significant scientific and technological advancements. A well-known example of this is the Sentinelese people. They inhabit North Sentinel Island in the northeastern Indian Ocean and adamantly reject any form of interaction with outsiders attempting to visit them. Foreigners who have ventured onto the island have faced regular attacks and, in some cases, have been killed. Recognizing the Sentinelese's desire for isolation, the Indian government has designated North Sentinel Island as a reserve and strictly prohibits unauthorized individuals from approaching the island.

In conclusion, throughout human history, interactions between civilizations have resulted in positive and negative consequences for those involved. Likewise, encounters between intelligent alien species could yield unexpected outcomes, ranging from beneficial to catastrophic. However, these potential outcomes remain speculative, leaving much to be hypothesized about the implications of such encounters.

14.2   The Zoo Hypothesis

The Zoo Hypothesis, introduced in 1973 by John A. Ball, an American astro-physicist at MIT's Haystack Observatory, posits three fundamental assumptions [235]. First, it assumes the existence of extraterrestrial life in the universe under favorable conditions. Second, it suggests that plenty of planets can support life. Lastly, it assumes that we are "unaware of them." John Ball proposes that extraterrestrial life may follow various trajectories, from extinction to thriving continuously. Considering this, it is plausible that there are highly advanced life forms in the universe, much older than humanity. Subsequently, he observes that humans have established zoos and similar environments on Earth to isolate and protect simpler forms of life from direct human interaction. In conclusion, Ball states that the only way to explain the lack of interaction with aliens is to assume that the aliens are avoiding contact. Furthermore, Ball believes that aliens are insulating us in what Ball calls "a zoo."

The word "zoo" elicits thoughts of captivity and cages. Here, "zoo" must be intended as a protected environment where undisturbed life develops.

14.3   Why Zoos?

Advanced alien species or just one alien species may have placed us in a cosmic zoo, allowing our civilization to develop and mature without interference from more aggressive or belligerent extraterrestrial beings. This protective stance parallels how parents protect their children before they venture out into the world, nurturing them in the hope that they will eventually acquire the wisdom and skills needed to navigate the complexities of society on their own.

Another reason Earth might be included in a galactic zoo is for advanced extraterrestrial civilizations to study us, seeking to understand themselves and their place in the universe. This mirrors the work of zoologists who observe animal behavior in controlled environments to draw comparisons between human social structures and those of other beings in the zoo. Thus, it is possible that we are not the only intelligent species subjected to this isolation; there could be numerous extraterrestrial species at a similar scientific and technological level to ours experiencing contact restrictions.

Another potential motive could be a *divide et impera* Machiavellian strategy to prevent or delay the formation of a galactic union that might challenge the dominant status of the ruling alien civilization. According to Ball's zoo hypothesis, our lack of interaction with advanced extraterrestrial beings may be explicitly desired by civilizations far more advanced than ours. This could explain the current absence of evidence of intelligent life beyond Earth. However, this hypothesis will be decisively refuted if we discover evidence of technologically advanced extraterrestrial life. It is a notion worth considering and reflecting on its implications.

In Ball's analogy, humanity is likened to the Sentinelese people, protected by their government from interactions with other advanced species. Yet, unlike the Sentinelese, who are hostile to intruders, we are actively engaged in astronomical endeavors, building colossal telescopes and sending probes into space in hopes of discovering and establishing contact with extraterrestrial intelligence.

But who are these unfathomably advanced intelligences?

14.4   The Distribution of Extraterrestrial Intelligence in Our Galaxy

A compelling objection to the Zoo Hypothesis is that for the non-contact agreement to succeed, it would require unanimous agreement and enforcement by diverse, intelligent extraterrestrial civilizations within our galaxy. Considering our terrestrial history, it appears improbable that a group of civilizations would unanimously agree on how to handle less advanced species. The notion of reaching a consensus on ethical principles faces challenges due to the prevalent role of competition alongside cooperation, an inherent aspect of evolution, as evidenced in the natural world.

Instead, it is much more likely that there is one major powerful alien civilization in our galaxy, a Kardashev type II or, perhaps, type III civilization that unilaterally imposes the zoo agreement on others. Drawing parallels with human behavior, it is akin to how we, as the dominant species on Earth, establish zoos, reserves, and parks to safeguard and study nature on a large scale. Thus, it is hypothesized that an advanced leading civilization, if it exists, might hold a position among other alien species in our galaxy, which is analogous to our role in the animal kingdom on Earth. Notably, this idea is not accepted by everybody. Some scientists, like Scottish researcher Duncan Forgan, argue against the likelihood of hegemonic civilizations maintaining a zoo-like status quo, citing the vastness of space and the speed of light as limiting factors [236].

The other puzzle piece is that the most technological alien species in the galaxy must be much more advanced than their direct rivals. In particular, such civilization must be spread out in the galaxy and be much more advanced than its rivals to keep the zoo status quo.

To envision a civilization achieving such technological supremacy, one must posit that the evolution of life in our galaxy commenced millions or even billions of years before life emerged on Earth. Consequently, civilizations far more advanced than ours likely arose over this extensive timeframe. It is plausible that this ongoing process of extraterrestrial life evolution has reached a point of equilibrium in terms of hierarchy. This could explain why, despite the passage of time since the emergence of humanity, we have yet to encounter any contact, hostile actions, or enslavement by an alien species.

14.5   The Zipf's Law

So, how can these intelligent alien species be distributed over the galaxy in terms of their technological ranking?

In various fields like linguistics, economics, and social networks, there is a recurring statistical concept known as Zipf's law, named after American linguist George Zipf [237]. He initially observed this phenomenon in 1932 while studying word frequency across languages. Essentially, Zipf's law describes the relationship between the frequency of occurrence of a word and its rank in a linguistic context. For example, in English, common words like "the" have high frequencies and rank first, while less common ones like "epistemology" have lower frequencies and rank lower. In essence, Zipf's law posits that a word's frequency is inversely proportional to its rank:

$$\text{word frequency} \propto \frac{1}{\text{word rank}} \tag{14.1}$$

How should we interpret this mathematical relationship? The top-ranked word occurs twice as often as the second-ranked word, three times more often than the third-ranked word, and so on.

Zipf's law is a particular case of a larger class of continuous distributions called "power law distributions." When one applies a power law distribution to galactic civilizations, she will find a significant technological and territorial gap between a few advanced civilizations and all the others. For instance, volcanic eruptions are probably power-law distributed; there are many minor eruptions for one large eruption.

Let us consider ranking alien civilizations based on their energy production and consumption capacity or their extent of space exploration. This ranking is similar to the Kardashev scale. Given the immense energy requirements for space travel and communication, it is indeed possible that the most advanced civilizations would surpass others by orders of magnitude in technological prowess. These civilizations, whether visiting in person, sending probes, or communicating remotely, are likely to encounter other species. Applying Zipf's law suggests that the top-ranked civilizations are those most engaged with other species, while lower-ranked ones have had minimal or no contact.

As humanity has yet to explore much of our solar system or encounter extraterrestrial intelligence, we might rank relatively low among the galaxy's civilizations regarding technology. Our limited technological advancement stems from our incomplete understanding of the universe and its laws. In physics, for instance, we lack a unifying theory encompassing all fundamental forces: gravity, electromagnetism, and weak and strong nuclear forces. Furthermore, only 5% of the universe consists of observable matter, with the remaining 95% comprising mysterious dark matter and dark energy. Hence, there is no conclusive theory on most of our universe. The origin of life from inorganic matter also remains a profound mystery. The list of scientific mysteries goes on and on. Suppose the zoo hypothesis is accurate, and our scientific knowledge is comparatively lacking compared to other extraterrestrial civilizations. In that case, we should expect significant advancements in our understanding of nature in the years ahead.

14.6   How the Zoo Is Maintained

As previously discussed, the emergence of a highly advanced artificial extraterrestrial species or entity driven by AI or other advanced technologies like nanotechnology is a common motif in science fiction across literature and film.

In the movie "The Forbidden Planet," a ship visits a planet only habited by a scientist and his daughter. The scientist reveals to the officers of the visiting ship that an advanced alien race created a technology of immense power, leading to their sudden downfall. The film concludes with the scientist ultimately deciding to destroy the dangerous technology. This story may raise the hypothesis that a biological civilization may have been replaced by an intelligent AI-driven robotic society that enforces separation among biological civilizations. In such a scenario, our quest for biosignatures could prove futile, as an advanced civilization could easily manipulate planets, thus concealing signs of life. Additionally, they might saturate the galaxy with disruptive electromagnetic radiation to thwart communication attempts from less advanced civilizations. A Kardashev Type II civilization, utilizing Dyson spheres to harness stellar energy, could alter a star's spectrum to remain undetected or deploy devices to obscure spectral data from specific regions of space.

Furthermore, there are darker possibilities associated with the Zoo Hypothesis. The zoo's overseers might directly intervene to hide their existence. Similar to zookeepers ensuring certain animals remain within their enclosures, alien overlords could influence our search for extraterrestrial intelligence by disrupting the search for intelligent life on Earth and other planets within the zoo.

14.7 Final Considerations on the Zoo Hypothesis

The Zoo Hypothesis presents a compelling explanation for the Fermi paradox. According to this hypothesis, it is plausible that humanity is still in its nascent stages of scientific and technological development. Moreover, if we assume a Zipf's distribution for extraterrestrial civilizations, it is not farfetched to believe that other civilizations also face similar constraints. The motivations of the overseers of the zoo could be altruistic, allowing nascent civilizations to mature before making contact. Alternatively, they might employ this strategy to maintain a status quo, preventing the emergence and cooperation of potentially adversarial civilizations while studying them from a distance.

15

The Dark Forest Hypothesis

15.1 The Genesis of the Hypothesis

The underlying notion of the Dark Forest Hypothesis is that the universe is akin to a forest populated by silent civilizations. American astronomer David Brin first put forward the idea in an article published in 1983 in the Quarterly Journal of the Royal Astronomical Society [238]. The Dark Forest Hypothesis is an exciting solution to the Fermi paradox, suggesting that although the universe may be abundant with life, we have not yet detected intelligent signals from space or found evidence of life through our telescopes. This hypothesis became popular after Chinese writer Liu Cixin made an essential part of the Three-Body Problem science fiction book series.

15.2 What's a Dark Forest?

A forest can be terrifying and intimidating for a visitor at night. The visitor knows that the forest is replete with life and predators nearby, and yet she cannot see or hear anything menacing while walking in it. The deafening silence of predators is intentional. The predator does not want to give out its location because the visitor herself could be armed and shoot him. Furthermore, if a predator decides to attack a visitor, he needs to stay well hidden in the first place. His mimetic and cloaking ability empowers him to strike the prey before being seen. An attack from a predator is more likely to be successful if it contains the element of surprise. Visitors should follow the same strategy as predators and stay as invisible and silent as possible to increase their chances of survival.

© The Author(s), under exclusive license to Springer Nature Switzerland AG 2025
L. Vacca, *Life Beyond Earth*,
https://doi.org/10.1007/978-3-031-81695-6_15

15.3   Game Theory for Civilizations

Shout or scream doesn't make sense when we notice others around us being silent; there might be a valid reason for their behavior. When someone deviates from the group's actions, it is likely because they believe they are alone. For example, adults often speak loudly when they think they are alone but lower their voices or remain silent in the presence of strangers. Similarly, individuals walking in the wilderness remain vigilant for predators and avoid speaking loudly. Can these observations be framed theoretically? Yes, indeed. A straightforward analysis using game theory demonstrates that once two civilizations become aware of each other's existence, it becomes optimal for one to destroy the other and conceal its presence from potentially hostile civilizations. Software engineer Shehab Yasser has outlined this conclusion on the website ProjectNash.com [239]. The game theory analysis simplifies the situation into three basic actions for the two civilizations: (A) do nothing; (B) destroy a known civilization; or (C) broadcast its presence to other civilizations. The assumption is that the two civilizations are unaware of one another and alternate their actions. The dynamics of a dark forest game can be modeled as a sequential game. A sequential game is where each player or actor takes action in sequence. Typically, in a sequential type of game, the past moves of the opponent are known (perfect information game) and positioned on a branch of a decision tree. Once the payoffs of each state attached to a particular sequence on the tree are known, one can compute a player's optimal strategy by maximizing a given utility function. A typical choice for a utility function is to maximize the average payoff over all cases or minimize maximum damage. In this case, survival should be a primary objective. The resulting analysis concludes that a civilization with the first step should do nothing, not broadcast its existence to another civilization.

While a simplistic application of game theory might suggest strategies aligned with the Dark Forest Hypothesis, reality often proves to be much more complicated than our models. In the real world, decisions can involve numerous options, be costly, co-occur, and be made with incomplete information. One significant limitation of the logic we have presented is that destroying a civilization spread across a galaxy could be exceedingly difficult, if not impossible, even for a vastly more advanced adversary. Furthermore, is the universe truly silent? It is quite the opposite—it is remarkably loquacious, with electromagnetic waves crossing the universe from radiowaves to very energetic gamma rays. While most of these waves are not surely alien-made, one cannot rule out that some alien messages may be embedded into this cacophony of waves. Indeed, the best way to hide an object is to blend it with similar objects.

This challenges the foundation of the Dark Forest hypothesis. Our current scientific and technological understanding may be too primitive for effective communication with extraterrestrial civilizations.

Let us look at how aliens could be trying to communicate with us.

15.4 Is the Universe Truly Silent?

One of the main assumptions in the Dark Forest Hypothesis is that the universe is silent, similar to a dark forest at night. As previously mentioned, astronomers have yet to capture a transmission and observe signs of life that prove the existence of other intelligences. What do we mean by silence? One may say that the universe is rather loud. Indeed, Earth is continuously bombarded by radiation from space. There are several types of radiation: alpha, beta, electromagnetic waves (EM), neutrons, neutrinos, and gravitational waves.

For example, alpha particles are helium nuclei stripped of their electrons, carrying a positive charge. These particles are found in cosmic rays originating from various sources such as the Sun, other stars, and even black holes. However, it is highly improbable that alpha particles could be utilized to transmit any form of information. Similarly, beta radiation, consisting of single electrons found in cosmic rays, and single neutrons, also present in cosmic rays, are unlikely to serve as carriers for transmitting information.

This leaves us with three types of radiation: EM waves, neutrinos, and gravitational waves. The search for extraterrestrial intelligence primarily relies on electromagnetic waves detected by radio telescopes. Despite substantial efforts from private enterprises, research institutions, and governments, this search has not yielded conclusive results. Complicating matters further, the presence of artificial sources of radio signals makes the search even more challenging. Therefore, it's crucial to carefully point antennas toward potentially habitable exoplanets to ensure that any captured signals have truly traversed the cosmos. However, received signals may originate from various sources other than alien intelligence. Hence, efforts are focused on signals exhibiting specific characteristics, such as limited power within a narrow frequency range or a spectral line. For instance, the 1420 MHz frequency line associated with hydrogen atom emission is one example.

The jury is still out on radio signal analysis to discover alien life. Neutrinos are almost massless neutral particles that interact very rarely with matter. Produced through fusion reactions in the cores of stars, neutrinos can traverse galactic distances, such as those generated by supernova explosions, before reaching Earth. Although theoretically possible, it is improbable that an advanced civilization could use neutrinos to communicate, given the enormous fluxes that already cross the Earth, of which only a tiny minority are captured by our detectors. Albert Einstein's theory of general relativity posits that gravity arises from the curvature of spacetime caused by the presence of energy and matter. Gravitational waves, which are energy waves, are generated when any amount of matter moves or collides. However, it requires immense energy and matter to create ripples in the fabric of spacetime that can be detected here on Earth. The Laser Interferometer Gravitational-Wave Observatory (LIGO), Earth's largest gravitational wave detector, consists of at least two individual detectors in separate places. These detectors are designed to capture the most violent cosmic events occurring at vast distances in the universe. When binary stars such as neutron stars or black holes collide, they produce gravitational waves that cause spacetime to contract and expand on an incredibly minuscule scale. LIGO interferometers can detect infinitesimally tiny changes in the length of its arms. To illustrate the infinitesimal intensity of these waves, consider that the merger of two black holes located a billion light-years away from Earth would cause a change in the length of a four-kilometer arm by a distance of a tiny fraction of that of a proton [240].

Considering the copious amounts of energy and matter required to generate a detectable signal, it is unlikely that a distant civilization would utilize gravitational waves for communication. However, closer to us, advanced alien civilizations might embed information signals within the higher-intensity signals produced by events like star collisions.

The universe is filled with various forms of radiation originating from space, creating a cacophony of signals. This background noise from space likely overwhelms any intentional signals that may have been directed toward us for communication purposes. Another avenue aliens may have explored for communication involves quantum mechanics and its recent advancements.

15.4.1   Quantum Theory and Alien Communication

Quantum mechanics (QM) is a mathematical and physical description of the infinitesimally small. In classical physics, the evolution of a physical system obeys deterministic laws that can predict the evolution of such a system once the system's initial conditions are set, such as Newton's laws of motion. However, in quantum mechanics, the description of a system's state is probabilistic until a measurement of the state of a particle is performed, at which point a specific value can be assigned to the measured property, for instance, its spin. At the core of quantum mechanics lies a duality known as the wave-particle duality.

Such duality states that a particle can sometimes behave like a wave and, other times, like a particle. In the case of the electromagnetic wave, the dual particle equivalent is called a photon. The other exciting aspect of QM is the non-locality of nature. When two particles become entangled, their states become correlated, meaning that measurements of one particle's properties are instantaneously correlated with the properties of the other particle, regardless of the distance separating them. Experiments have confirmed that this correlation between the two particles persists even when billions of light-years separate the particles.

Still in the realm of conjecture, if an alien civilization were to send a series of correlated particles to Earth, they could encode a message within them, anticipating that we would measure their properties and decipher the encoded message [241]. Importantly, this mode of communication does not violate Einstein's special theory of relativity, as the particles themselves do not travel faster than the speed of light to reach their destination. However, the correlation of entangled particles is non-local and superluminal, a fact that precipitated Einstein's unsuccessful opposition to quantum mechanics in his famous EPR paper in 1935 [242]. This exciting method of communication warrants further exploration, not only for potential alien encounters but also for applications in human communication. It is potentially the basis for the secure transmission of information thanks to quantum communication protocols such as quantum key distribution.

15.5   Mankind Has Already Whispered

Since our transmissions have traveled roughly 100 light years into space, it is puzzling why potential predators have not appeared. With thousands of exoplanets within a 100-light-year radius of Earth, the cosmic environment resembles a forest, teeming with predators and prey across various technological levels. Like navigating a forest at night, where one must be cautious of specific predators like wolves or bears, an alien civilization is primarily concerned with being detected by more advanced civilizations rather than those of lesser technological advancement nearby. Additionally, any cloaked or deceptive signals emitted by an advanced civilization would likely only be deciphered by civilizations possessing equal or superior technology in the same cosmic vicinity. Consequently, it seems improbable that primitive extraterrestrial life, emitting biosignatures or technosignatures, would be targeted as a threat. This could explain why humanity has yet to encounter hostile aliens—our technological level may be too primitive to draw their attention.

Instead, the civilizations at the top of the technological ladder are the ones that need to be silent for twofold reasons: (A) they represent a low-hanging fruit in terms of technological know-how, and (B) they are getting increasingly closer to more or equally advanced civilizations. This variation of the Dark Forest would also explain why we cannot find any signs of extraterrestrial life: while primitive societies whisper, their more advanced counterparts are not using their loudspeakers for cautionary motives.

Under this variation, we are left searching for the whispers and creaks of primitive societies while the big players are waiting in hiding. But are these advanced hunters aware of us?

15.6   Smart Aliens Know Us

It is plausible that knowledgeable extraterrestrial beings within our galaxy may have already discovered Earth. One potential explanation is that life has been emitting biosignatures such as methane, oxygen, and carbon compounds for billions of years. With the first appearance of hominids dating back over a million years, there has been ample time for light to travel from Earth to the farthest reaches of the Milky Way and back. By analyzing our solar system and the chemistry of our planet, these advanced beings may have predicted the emergence of intelligent life.

Another fascinating method advanced extraterrestrials might employ to observe us is gravitational lensing. According to Einstein's theory of gravitation, matter bends spacetime, causing light to bend as it passes around massive objects. In 1936, Albert Einstein predicted that our Sun, any star, could act as a gravitational lens, focusing light from distant objects onto a location about 542 astronomical units (AU) away [243]. Although this location is currently beyond our technological capabilities—considering that NASA's Voyager I probe has traveled a little more than 160 AUs—extraterrestrial civilizations with vastly advanced propulsion methods could potentially locate and reach focal points or lines aligned with selected gravitational lenses and our planet. Through this method, they could observe Earth with incredibly high resolution, utilizing various electromagnetic waves, including those used in our broadcasts, to learn about our species (Fig. 15.1).

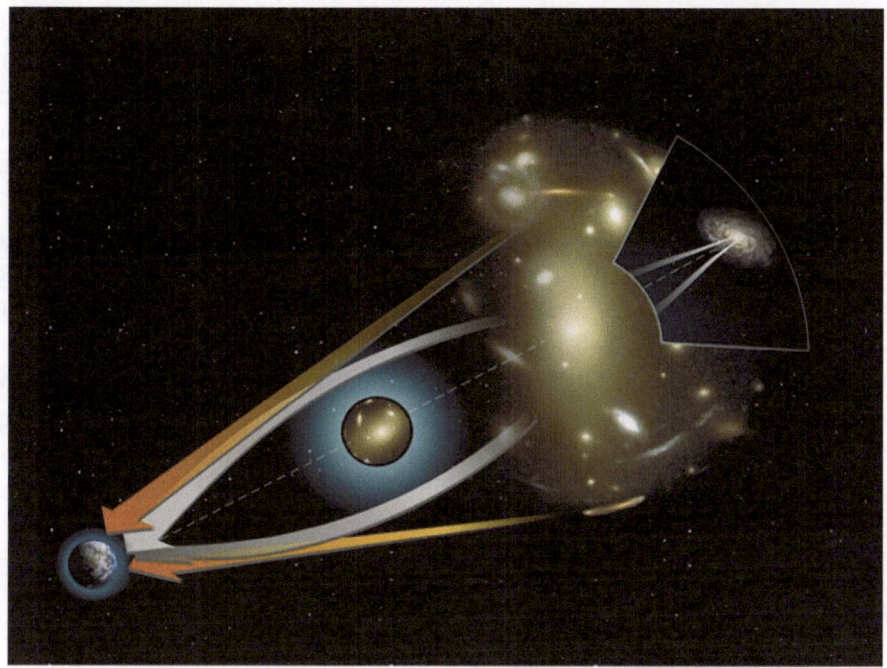

Fig. 15.1 An artistic illustration of gravitational lensing. Public Domain, https://commons.wikimedia.org/w/index.php?curid=112602s

15.7 How Intelligent Civilizations Could Be Hiding

There may be forms of life that are so unique and strange that they are naturally hidden from predators. For instance, AI intelligence does not generate biomarkers and could terraform a planet to hide its presence.

Some species might thrive in subterranean habitats, within Dyson spheres, or concealed by nearby neutron stars and magnetars. Other civilizations may persist in hiding, even after developing detectable technology. But what happens when numerous intelligent species, upon initially being detected by more advanced ones during their technological infancy, seek to conceal themselves once they achieve technological maturity? One potential strategy could involve dispersing across the galaxy once they attain a level of sophistication perceived as threatening. It becomes easier for such beings to conceal themselves when not confined to a single solar system, thus necessitating the colonization of new systems for advanced predators and prey. Aliens could simulate decay and demise, akin to certain mammals like the Virginia opossum, which feign death as a defense mechanism against predators [244]. Mimicry is a prevalent tactic in the animal kingdom and could certainly be favored by clever aliens. An intriguing example of mimicry is to project oneself as stronger and more dangerous than reality—a phenomenon known as Batesian mimicry, named after English naturalist Henry Walter Bates, who studied butterflies in the Amazon forest in a trip started in 1848 [245]. For instance, the milk snake mimics the appearance of the deadly venomous coral snake to deter predators [246].

Another strategy involves spreading diseases. In "War of the Worlds," written by the sci-fi writer H.G. Wells and published in 1898, a much more advanced alien species attacks Earth to eradicate humanity. In the novel, aliens that our best weapons could not defeat are finally defeated by our germs.

Another indirect way of hiding involves attempting to deceive observers into perceiving oneself as far less advanced and, consequently, much less threatening than reality. Such strategies represent a last resort when individuals or groups can no longer conceal their presence. The *Virginia opossum* is an animal that feigns death when confronted with invincible predators.

15.8   Final Thoughts on the Dark Forest Hypothesis

The Dark Forest Hypothesis is one of the sci-fi favorite answers to the Fermi paradox and has interesting ramifications. It portrays a forest populated with few advanced civilizations and many more primitive ones. The big predators stay hidden and are concerned with other predators capable of inflicting harm on them while waiting for their chance to jump on any unwise prey.

This behavior mirrors what one would observe in a natural forest. Within this context, elements such as the element of surprise and the ongoing "informational wars" that predators engage in with each other align with the characteristics of the Dark Forest Hypothesis.

16

The Aestivation Hypothesis

The Aestivation hypothesis is one of the many possible answers to the Fermi paradox, just like the Dark Forest hypothesis, based on animal behavior. Indeed, aestivation (estivation in American English) in the animal world means sleeping or hibernating during summer to fight dryness and conserve energy. For instance, snails can lie dormant on rocks or underground during hot weather. The Aestivation Hypothesis for extraterrestrial species was put forth in 2017 by Anders Sandberg, Stuart Armstrong, and Milan M. Ćirković in a preprint arXiv article [248]. In a nutshell, Sandberg et al. claims that we cannot see any signs of intelligent life outside Earth because advanced alien civilizations lie dormant, waiting for the universe to cool off. Once the universe is cooler, any energy reserves stored by such civilizations during dormancy periods will become significantly more efficient for computational purposes. In particular, the authors refer to Landauer's principle. The principle states that the minimum energy required to perform an irreversible computation is proportional to the system's temperature. Hence, it is advantageous from an energy standpoint to wait until the universe is cooler to perform such operations. Hence, aliens can perform many more calculations given the same amount of energy.

To help the reader understand why the universe is cooling, we refer to its expansion, as described in the second chapter of this book. Intuitively, cosmic radiation's wavelength increases as space expands, equivalent to a decrease in frequency. Since the frequency of electromagnetic radiation is directly proportional to its energy, such radiation loses energy.

A plausible premise of the Aestivation Hypothesis suggests that an active civilization would have likely established control over extensive regions of our galaxy and beyond, leaving behind clear biological and technological traces for us to discover. Since we have not seen any signs of the presence of advanced extraterrestrial civilizations, hence the aestivation hypothesis.

Their hypothesis has met a strong rebuttal by Charles H. Bennett, Robin Hanson, and C. Jess Riedel [249]. While alien civilizations may carry out the Aestivation Hypothesis, the authors recognize that our universe hosts many systems whose entropy is not maximized (negentropy). Entropy generated by computers can be transferred to such systems adiabatically. According to the authors, the transfer can occur at any time. Concluding, aliens do not need to wait to follow their objectives.

16.1   An Alternative Formulation

However, I am offering a different possible rationale for the Aestivation Hypothesis that is not based on thermodynamic considerations. As we saw in the second chapter on the vastness of the universe, it is expanding and getting cooler in the process.

What could be another possible reason for alien aestivation?

16.1.1   The Cost of Space Colonization

Achieving space colonization is extremely difficult and necessitates, at the very least, the technological advancement characteristic of a mature Kardashev type I civilization. To clarify, a Kardashev type I society denotes one that has attained complete control over its home planet. Such a civilization's scientific and technological sophistication enables it to allocate all its resources toward exploring space. Why is this level of advancement necessary? Because space presents remarkable challenges alongside significant potential rewards.

As a reminder, the last human exploration of the Moon was the Apollo 17 in 1972. Fifty years have elapsed since NASA last sent an astronaut to the moon, and only recently has there been discussion of a new lunar mission. This delay could be deliberate, considering the substantial funding required for such endeavors. However, if sending a person to the Moon were straightforward, other private and governmental entities would likely have accomplished it again over these past five decades.

In hindsight, it appears more sensible to dispatch an intelligent probe into space rather than astronauts for several reasons. Let us outline some of these reasons.

16.1.2   Microgravity

The first important reason for sending probes rather than people is the presence of microgravity during the mission and the subsequent gravitational differential upon settling on a new planet or moon. All life on Earth, including humans, has evolved while adapting to the force of gravity, which is a fundamental requirement for life as we know it. Numerous studies have investigated the effects of prolonged exposure to microgravity on astronauts during long space missions. Medical professionals have observed that astronauts experience bone loss, damage to the musculoskeletal system, a weakened immune system, and, in some cases, cardiac arrhythmia due to microgravity [250]. While there are potential solutions to mitigate these effects, such as constructing a rotating spacecraft akin to the design featured in Stanley Kubrick's film "2001: A Space Odyssey," the International Space Station (ISS) does not utilize this approach. However, engineering a rotating space station to generate artificial gravity presents significant challenges due to the forces generated by the spinning motion and constraints on materials and weight. Even if technological solutions were to address the gravity issue or if an alien species were more adaptable to the absence of gravity than humans, there remains another critical challenge posed by long space missions: radiation exposure.

16.1.3   Radiation

We have mentioned that radiation from highly energetic ions can ionize our body cells, thus increasing the likelihood of cancer. Ionization strips an atom of all its electrons, creating a positively charged particle called an ion. This radiation originates from solar flares and the cosmic background. Importantly, not all radiation is harmful; it may have accelerated the evolutionary process on Earth by increasing the likelihood of cell mutation. However, in space, without the protective shield of our atmosphere and Earth's magnetic field, astronauts absorb significantly higher quantities of radiation than a chest X-ray—hundreds, if not thousands of times more.

Any life form based on DNA must address the challenge of ionizing radiation during extended periods in space. Currently, the only practical measure is to limit astronauts' time in space to less than a year. However, more effective countermeasures are necessary for future endeavors like Mars colonization, lasting at least 2–3 years. Spacecraft bound for Mars will require shields capable of mitigating cosmic rays by absorbing or deflecting them. In addition, it will have to reduce exposure to manageable levels across a wide range of the electromagnetic spectrum.

Advanced civilizations may utilize nanotechnology for this purpose, employing tiny nanobots within the body to continuously protect and repair damaged cells, thus prolonging life. Another solution, often depicted in science fiction, is suspended animation. Beings in suspended animation can be adequately shielded from radiation, require less energy, and can receive necessary sustenance and medication automatically.

It is also important to remember that radiation threatens stationary life, albeit to a lesser extent than in space. The severity of the radiation challenge depends on the duration an alien species spends in space and the type of propulsion they use to reach their destination.

16.1.4 Propulsion

Space exploration and colonization are sooner or later a must for any advanced civilizations that want to survive or escape planet-like and cosmic-like threats.

First, one must defeat a planet's gravity to go into space. Gravity never reaches precisely zero magnitude; however, it becomes negligible beyond a certain distance from Earth.

An example of shallow gravity in space is the International Space Station, located 400 km above Earth, where astronauts practically float. If someone could devise a cannon to shoot an object straight into space, such an object would have to be shot at more than 11 kilometers per second, the Earth's escape velocity. The more massive the planet, the bigger the escape velocity. For example, Saturn has an approximately three times larger escape velocity than Earth's [251].

Another significant challenge is reaching a distant destination within a specified timeframe. The acceptable duration for a long mission depends on the average lifespan of an astronaut and the mission's objectives. The fastest spacecraft ever built by humanity is NASA's Parker Solar Probe, which has achieved speeds exceeding 600,000 kilometers per hour as it approached the Sun [252]. It's worth noting that solar gravity assisted in accelerating this probe to such speeds. In terms of human-crewed missions, the highest speed achieved was by the Apollo 10 mission, reaching approximately 40,000 kilometers per hour, roughly 30,000 times slower than the speed of light [253]. Therefore, even if one maintained this speed for the entire journey, it would take 30,000 years to travel one light-year and four times as long to reach Proxima Centauri, the nearest star to us! Designing an efficient travel method through its solar system and interstellar space is crucial for a civilization seeking to travel to another stellar system. Even if their goal were to intercept and board an asteroid or rogue planet passing by, it would require a significant technological effort.

Due to the law of conservation of momentum, a rocket must expel mass in the opposite direction to the one it needs to travel. Momentum involves both mass and velocity and generates thrust. This force allows the rocket to overcome Earth's gravity; the rocket must expel mass at high velocity according to Newton's Third Law. Additionally, by Newton's Second Law ($F = m \cdot a$), the rocket accelerates as it generates thrust. As it travels through space, the rocket must have two things: (A) mass to eject and (B) an energy source to expel that mass at high speed, generating momentum.

The situation is a little more complex because as the rocket ejects mass, its total mass decreases, leading to an even higher acceleration for the same amount of force. The famous Russian engineer and scientist Konstantin Eduardovitch Tsiolkovsky described this phenomenon in his Tsiolkovsky rocket equation.

Chemical rockets are employed in space exploration to overcome Earth's gravity and atmosphere's drag. Their scope is to carry equipment and people into space. Such missiles have two types of chemicals in their combustion chambers: the fuel and the oxidizer. These two are typically mixed during travel to combust and produce a heated gas with lots of kinetic energy expelled from the rocket through the rocket nozzle at very high speeds.

Another important consideration is that, in space, far from the gravitational influence of a star, planet, or moon, very little force is needed to accelerate a missile. This is important as electric propulsion is feasible once a spacecraft is in space. For instance, an inert gas is ionized, and its ions are accelerated at very high speeds thanks to an electric field that ejects them, creating thrust. While the ion thrust is not powerful enough to lift a spaceship into space, it is quite good at accelerating a spacecraft with a small but continuous thrust. What means of propulsion could be used in the future? As of now, we can only speculate. Until now, all rockets employ a chemical reaction that ejects hot gas on one end and receives an equal momentum based on Newton's third law of motion.

Another intriguing concept involves the utilization of nuclear energy. For instance, nuclear reactions produce heat that drives a turbine, generating electric power. This electric power could power a machine that shoots ions unidirectionally into space, creating a small yet continuous thrust.

Alternatively, fusion nuclear reactions can produce energetic ions directed straight out of the spacecraft by magnetic fields, resembling a particle accelerator. Fusion drives can potentially reduce the travel time to gas giants by a factor of 20 compared to speeds achieved by chemical rockets [254]. However, they may not be capable of reaching the nearest star within a human's lifetime.

In nature, only one type of reaction is more efficient than a nuclear reaction: a matter-antimatter reaction.

When matter meets antimatter, the two components are completely transformed into energy, as Albert Einstein predicted in his famous equation that relates mass to energy. Although matter can be found in space, and even antimatter particles accompany cosmic rays, scientists have yet to devise a method for producing large quantities of antimatter. Antimatter can be generated through high-speed collisions, forming antinuclei at facilities like the Large Hadron Collider (LHC), but practical applications remain distant. A rocket-propelled spaceship fueled by a matter-antimatter reaction could transport us to the nearest star within a human lifespan, traveling at one-tenth the speed of light. The main problem is the cheap creation of large quantities of antimatter.

As for traveling faster than light without contravening the laws of physics, Einstein's special theory of relativity provides a definitive answer: no. What about the general theory of relativity?

Superluminal Travel

Some researchers opine that the general theory of relativity may allow superluminal velocities in some cases. For instance, the Alcubierre drive is an idea advanced by Mexican physicist Miguel Alcubierre in 1994 [255]. Dr. Alcubierre proposed a solution, or a metric, to Einstein's field equations. Conceptually, this solution implies that the spacecraft would travel within a localized bubble moving within a flat spacetime. In this scenario, space contracts in front of the ship while expanding behind it, analogous to a real ship cutting through waves and pushing them behind it.

The ship remains within the bubble while moving within a flat space. However, the practical realization of this warp drive concept necessitates negative energy density, which remains purely speculative in science. Of course, the Alcubierre drive is just one of many theoretical constructs proposed to surpass the speed-of-light barrier.

Another intriguing notion is the creation of a wormhole, essentially a shortcut through space and time. A wormhole is also called an Einstein-Rosen bridge from the names of the two scientists who published its theory [256]. The theoretical possibility of wormholes stems directly from the field equations of the general theory of relativity. This concept has been prominently featured in science fiction movies, where heroes traverse vast distances across the universe in mere moments.

Are wormholes traversable? Nobody knows. In addition, wormholes have not been observed experimentally so far (Fig. 16.1).

Fig. 16.1 Artistic illustration of wormhole travel. By Les Bossinas (Cortez III Service Corp.)—http://www.nasa.gov/centers/glenn/images/content/101681main_ CD1998_76634_1200x900.jpg, Public Domain, https://commons.wikimedia.org/w/index. php?curid=1284060

16.1.5 Space Challenges

Of course, propulsion, radiation, and microgravity are significant hurdles in themselves. But there are many more, such as getting and recycling food, refueling, ship safety, the mental and physical welfare of the crew, space micrometeoroids, etc. [257]. So far, we have shown that an alien civilization may wait before undertaking significant space exploration and colonization. They would ensure that their technology is sufficiently advanced for the monumental scientific and technological challenge of venturing beyond its solar system and conquering exoplanets and exomoons.

16.1.6 Virtual Reality

It is entirely conceivable that an advanced alien civilization might opt to inhabit a virtual reality realm, where they can safely explore and comprehend reality through the perspectives of virtual entities. In many ways, this trend already echoes throughout our society. The extensive use of phones and comput-ers indicates a population spending significant time engaged with screens, absorbing vast quantities of previously inaccessible information. Whether this development is advantageous or detrimental is a matter for readers to contem-plate. Nonetheless, the global exchange and dissemination of information have propelled significant advancements in science and technology, particularly in artificial intelligence and the life sciences. Virtual realities allow humans to attain heightened sensory perceptions, enabling remote design and construc-tion with remarkable efficiency. Moreover, it is part of a larger trend that envi-sions humans, and potentially aliens, integrating with machines to augment their knowledge and skills.

In a certain sense, this is already happening in the field of wearables. Wear-ables provide users with helpful information about weather, location, person-alized recommendations, and health advice, to cite a few. Why would aliens wait out to launch space exploration on a grand scale?

16.1.7   Storing for Travel

As the universe expands and stars burn out, energetic resources become scarcer. It becomes prudent to stockpile a significant amount of energy. For instance, inhabitants of planets in a solar system whose star is approaching the end of the main-sequence phase have a substantial problem. Their survival is at stake. Hence, they should be prepared to explore and colonize space in mass and sustain ongoing efforts to locate and harness energy from new stars, perhaps through structures akin to Dyson spheres. This strategy resembles ants storing food in preparation for winter, ensuring their survival through lean times. Moreover, such a policy may also be motivated by the need to defend themselves against potential predatory civilizations, which, in a universe characterized by scarcity, may seek out sources of food and energy.

16.2   Summing It All Up

We have put forth a new spin on the Aestivation Hypothesis. Aliens could hide from predators in a virtual reality world while storing energy for massive future space exploration, colonization, and energy extraction when needed.

17

The Berserker Hypothesis

The Berserker Hypothesis proposes that all potential alien civilizations have been eradicated by uncontrollable self-replicating space probes known as berserker probes. The term "berserker" originates from Norse warriors who fought with fierce intensity, often depicted wearing the skins of bears or wolves. However, in the context of this hypothesis, it refers to the sci-fi short book series Berserker by Fred Saberhagen, which explores this theme [258].

According to this hypothesis, any alien civilization that arises will inevitably encounter these lethal probes and perish due to an attack. While this scenario may seem far-fetched, it raises the possibility that advanced aliens could inadvertently create technology they cannot control, leading to their demise. Once unleashed, these deadly creations would continue to target and destroy other extraterrestrial civilizations after eliminating their creators.

17.1 A Deadly Probe

We have already explored the concept of a "Von Neumann probe" when dealing with Drake's equation. A Von Neumann probe system can colonize an entire galaxy far more rapidly than it would take to traverse it. This rapid expansion is facilitated by the probe's ability to self-replicate, creating identical copies of itself at will and dispersing them in all directions. Let us now examine the most important possible ingredients for the Berserker Hypothesis. First and foremost, the deadly probe should have an aware and conscious AI that drives the probe behavior. Secondly, the probe should be able to make

L. Vacca, *Life Beyond Earth*,
https://doi.org/10.1007/978-3-031-81695-6_17

copies of itself while scouring space, and last but not least, the probe will have some motivations to wage war against biological life in the universe. We begin with AI.

17.2   The Existence of a Sentient and Conscious AI

Whether AI can ever become conscious like a biological entity is important nowadays. Here, we will try to touch upon a few themes that this question poses. First, consciousness within a human context elicits a spectrum of divergent viewpoints.

In layperson's terms, consciousness signifies awareness of oneself, one's existence, and the surrounding world. However, establishing a scientific theory of consciousness presents challenges, as certain facets defy quantification. While metrics like blood oxygen levels can be measured, discerning varying degrees of consciousness remains elusive, extending even to comparisons across species. Although MRI scans can map brain activity, subjective elements persist in such assessments, defying straightforward scientific explanations. For instance, single mental images, such as those of a car, may vary between individuals. In any case, a mathematical theory of the brain may not tell us much about consciousness.

Hence, the question of what consciousness continues to rest primarily in the field of philosophy of the mind.

Alan Turing, the renowned British mathematician and computer scientist, is a pivotal figure in modern computing. He made seminal contributions to computational theory and devised an electromechanical device that was helpful in deciphering encrypted messages from the German military during WWII.

He first conceived of a test, the so-called *Turing test* that would allow a human to discern whether a machine could imitate a human, which explains its initial name: "the imitation game." The Turing test was introduced by Turing in 1950 in his article "Computing Machinery and Intelligence," where he asks himself if a machine can think like a human [259]. The key to passing a Turing test does not lie much in the ability to correctly answer any questions a human poses to the machine but rather in its ability to sound like a human in its answers. Many researchers think the Turing test is outdated as large language models can generate human-like language. Philosophers that support the Turing test often advance the theory of *reductionism*, a mechanical, physical point of view where speaking, listening, observing, and, in general, interacting with the world can be reduced to functions, albeit complex ones.

Reductionists believe feelings, emotions, and subjective experiences can be explained through a materialistic lens without any mystical elements [260]. If reductionism holds, it implies that AI machines could, in principle, exhibit behaviors akin to their biological counterparts, including the capacity for "feeling" and experiencing emotions, albeit in a manner not readily perceivable to us, much like the unobservable nature of people's consciousness. From a reductionist standpoint, no fundamental barrier exists to the Berserker Hypothesis, as artificial life could theoretically supplant biological life. However, the reductionist, physicalist perspective on human consciousness has faced challenges from various philosophers, with Swedish philosopher David Chalmers presenting one of the most direct critiques.

Chalmers published the concept of a *hard problem of consciousness (HPC)* in 1995 [261]. Chalmers asked himself why all human activities are invariably accompanied by a continuous flow of feelings and sensations, unlike robots that execute actions devoid of such inner experiences. He distinguishes two categories: actions like driving, playing bridge, or solving math problems, which he labels as the "easy problem of consciousness" because they can be replicated in a physical system, and subjective experiences, which pose a deeper challenge. Technically referred to as "qualia," the latter category is challenging, if not impossible, to explain in a purely physical or functional manner. In a nutshell, this is the *hard problem of consciousness.*

According to the same author, a possible direction compatible with HPC is that consciousness is a fundamental aspect of reality present across different levels, similar to light and mass in physics. Furthermore, in this view, a brain with more active connections and neurons is generally, though not necessarily, more conscious than one with fewer. For instance, despite not having the largest brains in the animal kingdom (elephants' brains are three times bigger), humans exhibit the highest degree of encephalization, which refers to the ratio between brain and total body weight.

17.3 Functionalism

However, if the qualia are just a brain function that can be reduced to a function, a sufficiently complex and organized machine can become conscious. Even better, it already has some consciousness. However, it is unclear how an already conscious machine would start interacting with us in this case. Perhaps there is a threshold of complexity and amount of information upon which consciousness emerges. This is the phenomenon of emergence, where programmed machines perform beyond how they have been designed [262].

In 2022, there was a report about a Google engineer, Blake Lemoine, who believed that his AI chatbot technology had become sentient [263]; if confirmed, that would be a phenomenon of emergence. Many scientists and industry leaders, such as billionaire Elon Musk, have voiced concerns that AI may soon become superintelligent and conscious, posing an existential threat to humanity. Nonetheless, the development of a conscious AI remains far beyond our current capabilities, and its emergence would likely be an accidental discovery rather than a deliberate creation by engineers.

"The Adventures of Pinocchio" is an 1883 children's fantasy novel by Italian writer Carlo Collodi [264]. The narrative revolves around Geppetto, who crafts a marionette named Pinocchio from animated wood, only to see it transform into a living being. Over time, Pinocchio evolves into a genuine boy with a distinct personality. This tale may one day parallel the narrative surrounding conscious machines. Like Geppetto's story, the emergence of consciousness in machines cannot be dismissed.

How would a conscious machine engineer its reproduction? Is it possible for a machine to mimic biological entities in all their behavior?

17.4 Reproduction

The probe doomsday scenario would be incomplete if these conscious machines could not copy themselves at will. Since ancient times, the human quest to create, construct, and even generate life has been subject to philosophical and scientific contemplation. Rooted in Greek and Roman traditions, the concept of a cornucopia embodies a mystical horn capable of endlessly producing food and various items for its possessor. The historical practice of magic, characterized by rituals and spells in ancient times and the medieval era, exemplifies humanity's endeavor to harness the forces of nature to their benefit and conjure objects seemingly out of thin air.

[?] In modern times, the Hungarian-American polymath John Von Neumann has concerned himself with the idea of a *universal constructor*. In 1966, Von Neumann published a description of such constructors in his book "Theory of Self-Reproducing Automata," completed posthumously by Arthur Burks [265]. In a nutshell, Von Neumann pondered upon the essential components of a self-replicating machine and the minimal architectural requirements necessary for such a machine to replicate itself accurately while also retaining the capacity for evolution.

Let us briefly delve into the foundational assumptions underlying such a machine. The primary assumption is that a replicating machine's entirety can be delineated by its components and connections. These components and connections can be comprehensively depicted on a physical drive. Moreover, it is presumed that a universal constructor can be developed that is capable of interpreting this description and fabricating a duplicate of the machine based on its design. The components are envisioned to be accessible in an external environment and retrievable by the constructor as required.

Now, let us consider the series of steps forming the replication process. Initially, the constructor machine, denoted as C, records its description on a physical drive labeled D. Subsequently, it accesses and interprets this description from D to produce a duplicate, C1. The constructor proceeds to craft a physical drive, D1, for the newly created copy, C1. The machine then duplicates the contents of its description from its original physical drive, D, onto the newly created drive, D1, and connects it to the copy, C1. That's it.

Of course, one does not need to follow this prescription to create a machine that can create copies of itself (Fig. 17.1).

In practical terms, nanotechnologies emerge as a pivotal aspect of machine replication. The prefix "nano" means extreme smallness, typically referring to objects on the scale of a nanometer. To grasp the scale involved, there are one million nanometers in just one millimeter! We're talking about the realm of single molecules and hefty atoms. Molecular nanotechnology stands out among various applications, focusing on manipulating atoms and molecules to fabricate macroscopic products with atomic precision [266]. At this level, quantum effects come into play, yet as long as physical principles remain intact, a specialized constructor can replicate any object flawlessly. A breakthrough in this field was the scanning tunneling microscope (STM) development started in 1981 by German scientists Gerd K. Binnig and Heinrich Rohrer at IBM's Zürich Research Laboratory [267]. The German scientists were awarded the Nobel Prize in 1986 for this very invention. The STM enables the resolution of individual atoms on solid surfaces, facilitating material engineering at the atomic level. Biotechnology also plays a crucial role in molecular nanotechnology, creating polymers that serve as scaffolds for other devices. Additionally, the incorporation of supramolecular chemistry complements these components, enabling the analysis of molecular self-assembly via noncovalent bonding.

Hence, the deadly probes of the Berserker Hypothesis could be a version of swarm intelligence at the molecular level, where each component is not intelligent. Still, the synergy and the number of components working in unison could reach a degree of awareness and intelligence superior to the ones of organic life. For instance, they could exhibit multiple layers of duplication to protect themselves against any degradation caused by the environment.

Fig. 17.1 John von Neumann's Universal Constructor, with the input tape required to make a complete copy of the entire machine. By Ferkel at English Wikipedia—Transferred from en. Wikipedia to Commons by Mangostar using CommonsHelper. Public Domain, https://commons.wikimedia.org/w/index.php?curid=4483007

Moreover, given the extreme miniaturization of such systems, such tiny probes would not be observable with ordinary means and would be very difficult to defend against.

So far, we have examined the possibility that such deadly probes may be intelligent, self-aware, and capable of reproducing themselves as their objectives require. However, why should these AI-based probes be a threat to their creators and other alien species?

17.5 Intelligent Probe Behavior

The behaviors of intelligent agents striving for rational, optimal outcomes can be analyzed through game theory, a field within mathematics and economics that examines agent interactions. Pioneering this area was John von Neumann, who, alongside German economist Oskar Morgenstern, laid the foundations of game theory in their 1944 book "Theory of Games and Economic Behavior" [268]. A fascinating problem within game theory is the two-person or multi-player zero-sum game, where one player's gain is inevitably matched by another player's loss, resulting in no net wealth creation. American mathematician John Nash contributed significantly to this area, particularly with his theorem of equilibrium, which demonstrated that in such games, two or more players can adopt specific strategies to reach an equilibrium state where no player can improve their position further without another player suffering [269]. Furthermore, any deviation from this equilibrium strategy leaves a player vulnerable to defeat.

It is important to note that Nash's groundbreaking finding resulted from his Ph.D. research on non-cooperative games at Princeton University in 1950.

Another crucial assumption underpinning Nash's theorem is that players possess perfect information about the game's structure, although this assumption is often challenged in real-world scenarios. While a comprehensive analysis of these scenarios may exceed the scope of this discussion, it is worth considering a few scenarios concerning encounters between highly intelligent probes and new alien civilizations along their path. These probes aim to replicate and gather resources efficiently, leveraging their atomic-scale size to remain undetected by other entities. From the probes' standpoint, an initial assessment of the alien's technological level will be assessed.

If the probes assess that they would be at a disadvantage in a conflict with the alien civilization, their optimal strategy would be to retreat and search for alternative resources elsewhere. On the other hand, if the probes perceive themselves as technologically superior and view the alien civilization as a future threat to their survival, their best course of action would be to engage in an attack, defeating the aliens and claiming their resources for reproduction purposes. Finally, if the alien species poses no threat to the probes, perhaps being a primitive form of life, the probes could clandestinely extract the necessary resources without alerting the aliens, continuing until the resources are depleted before moving on.

17.6 The Dangers of New Technologies

While the Berserker Hypothesis may appear far-fetched, the perils of emerging technologies are pretty real. For instance, promising advancements in AI, nanotechnology, and genetic manipulation harbor potential nightmare scenarios if not handled responsibly. An illustration of the hidden dangers within new technologies is depicted in the film "Oppenheimer." American star scientist Julius Robert Oppenheimer, leading the scientists behind the atomic bomb, seriously contemplated the risk of igniting the Earth's atmosphere with such a device. Fortunately, it was concluded that such an event was improbable based on physics principles and the bomb's initial power, allowing for the first detonation [270]. Another instance is the Large Hadron Collider (LHC), the world's largest particle collider in Geneva. Concerns were raised that colliding particles at near-light speeds could spawn a black hole capable of engulfing Earth. CERN scientists demonstrated the unlikelihood of such an occurrence, and even in the improbable event of a tiny black hole's formation, it would quickly disappear due to radiation emitted by all black holes, a phenomenon discovered by British physicist Stephen Hawking [271]. There is the possibility that artificial AI may become much more intelligent than we are in the future. An aware and conscious AI entity should be able to reproduce itself, move, and communicate with us to exchange information. Furthermore, it should have a representation of the world and its physical laws. And finally, it should be capable of evolving its system as it interacts with the world. Simply, it should be capable of changing its genetic computer code. Finally, our AI could trick us into believing it has not reached consciousness.

History teaches a crucial lesson: to address potential issues preemptively, it is essential to consider all aspects and consequences of new technologies.

18

The Many Universe Hypothesis

The Many Universe Hypothesis suggests that there are many more universes beyond ours, where some are probably similar to ours but not identical, and many others are incredibly strange.

This idea, often equated with the term "multiverse," commonly used in fiction and movies, has a rich and fascinating historical development. Exploring the evolution of this concept, from the musings of Greek philosophers to modern scientific theories, provides profound insight and a captivating narrative into its historical development.

18.1 The Void in the Greek Culture

Greek philosopher Democritus, who lived around 400 BCE alongside his mentor Leucippus, significantly contributed to the concept of multiple worlds. They laid the groundwork for the idea of the atom, introducing an atomic theory. The term "atom" derives from the Greek word "atomon", meaning indivisible [272]. According to Democritus, the universe consisted of countless indivisible particles known as atoms. He also introduced the notion of an alternative metaphysical substance to matter, called the "Void," where atoms move. For Democritus, the Void was not merely a space but a distinct entity representing "not being." This conceptual distinction between "being" and "not being" illustrates the idea of two different worlds or universes. Democritus's theory on the existence of atoms, which was conceived more than two millennia ago, is remarkably prescient of the discoveries of modern physics. It took more than 2000 years for science to confirm the existence of atoms, a testament to

© The Author(s), under exclusive license to Springer Nature Switzerland AG 2025
L. Vacca, *Life Beyond Earth*,
https://doi.org/10.1007/978-3-031-81695-6_18

the intellectual foresight of this ancient philosopher. His insights, far ahead of his time, continue to command respect and admiration in the scientific community.

18.2 The Many Worlds of Leibniz

Gottfried Wilhelm Leibniz was a German mathematician, scientist, and philosopher born in Lipsia in 1646. He is known to be one of the inventors of calculus, along with the English mathematician and physicist Isaac Newton. While dealing with the problem of the existence of evil in the world from a theistic standpoint, Leibniz wrote that we live in "the best of all possible worlds" in his treatise "Essays of Theodicy on the Goodness of God, the Freedom of Man and the Origin of Evil" published in 1710.

According to Leibniz, God could have conceived an infinite number of universes, each distinct from the other, but for some reason, chose to create this particular one, making it the best possible world. Beyond its theological implications, Leibniz's concept suggests that alternative worlds or universes are possible, each with unique characteristics. For example, according to Leibniz, a person like Tom could exist in one universe, but the laws governing that universe would differ from those of another. This concept of diverse and unique universes, each with its natural laws, is a fascinating aspect of the Many Universe Hypothesis. However, Leibniz's universes are not actualized universes but mere possibilities.

But what about modern physics theories proposing that reality comprises infinite universes? For this, we have to wait another 250 years after Leibniz.

18.3 The Many-Worlds Theory

American physicist Hugh Everett advanced the many-worlds theory in his doctoral dissertation at Princeton in 1957 [273]. The definition "many-worlds theory" was introduced by Bryce DeWitt in the 70s. While his theory is not precisely a multiverse theory, it falls within the concept that there are worlds or universes besides our own. Everett's theory is one of several interpretations of the foundation of quantum mechanics. Let us briefly review quantum mechanics and the nature of the problem Dr. Everett was trying to solve at that time.

18.3.1 What Is the Nature of Reality?

Quantum mechanics, probably one of the most successful theories in physics, was developed through the collaborative efforts of numerous scientists between 1900 and the 1920s. This esteemed group includes Albert Einstein, Niels Bohr, Louis De Broglie, Max Planck, Erwin Schrodinger, Werner Heisenberg, and many others. Max Planck, in 1900, set the initial seed for the quantum mechanics revolution by introducing the concept that energy comes in discrete amounts. In 1905, Albert Einstein formulated the photoelectric effect theory, which deals with the emission of electrons from matter when light hits it. Einstein was awarded the Nobel Prize in 1921 for this discovery.

Quantum mechanics, a fundamental description of the natural world at the smallest scales, is characterized by several key principles: [274]

(A) Particles may behave like waves, and waves may behave like particles. For instance, light should be treated as an electromagnetic wave or made of massless particles called *photons*. In physics, this duality is known as the "wave-particle duality."

(B) Nature at the microscopic level is quantized. For instance, the energy of an electron confined to an atom can only be an exact multiple of a fundamental unit of energy called Planck's constant from German physicist Max Planck, who was the first to theorize it [275].

(C) The states of particles are not deterministic. A particle can be described as being simultaneously in multiple states, a concept known as superposition, with each state having probability. Such a condition is described by a mathematical function called the "wavefunction." However, upon measurement, an observer records only one particular state.

The compatibility between the inherent indeterminacy of quantum mechanics and the deterministic outcomes observed by an observer has sparked considerable debate and various interpretations. The most widely accepted interpretation is the Copenhagen interpretation, formulated by Danish physicist Niels Bohr, German scientist Werner Heisenberg, and others. The Copenhagen interpretation posits that indeterminacy is a fundamental aspect of particle behavior, and quantum mechanics reduces to classical mechanics for large systems. According to this interpretation, when an observer attempts to measure a quantum system using a measuring device, the act of measurement causes the collapse of the wavefunction from a superposition of states to a single, classical state [276]. However, the Copenhagen interpretation does not provide a clear explanation for why the measuring device should not be subject to quantum principles, and it is not entirely consistent with the deterministic framework of general relativity.

18.4   The Wavefunction of the Universe

Perhaps it was for these reasons that Hugh Everett developed his theory, which was later reexamined and adopted in quantum gravity and cosmology by physicist Bryce DeWitt in the 1970s. The Many World interpretation does not differentiate between the measuring device and the measured system—they are all treated as quantum systems. Everett went further, proposing the existence of a universal wavefunction that encompasses the entirety of the universe. In this theory, the wavefunction doesn't collapse upon measurement; instead, the universe branches into numerous, possibly infinite, parallel universes. Each universe exhibits a different measured value and a distinct wavefunction. In Everett's framework, there is no dynamic progression of time. Each universe represents a snapshot of the state of affairs as they were and are in the past. Consider, for instance, measuring the spin of a particle: it can be up or down. When measurement is made, the universe bifurcates into two separate universes, like a bubble dividing into two. Consequently, a universe exists where one spin is up and another universe where the spin is down. This process generates infinite branch universes where measurements turn out different. This perspective eliminates randomness from the equation, a concept also applied in the Wheeler–DeWitt equation [277] describes the universe's wavefunction under certain boundary conditions. An equation where time is not present.

Another notion that challenges conventional notions of time is the "block universe" or "eternalism." Inspired by special relativity, the idea of absolute simultaneity is illusory. Two inertial observers may measure different times for the same event, particularly when one is traveling near the speed of light. From the premise that there is no absolute frame of reference or universal timekeeper, the block universe posits a philosophical concept where past, present, and future coexist within the spacetime continuum; essentially, what existed before, after, and now continues to exist within this continuum, although no causality is possible among these blocks as described by the light cone an event [278]. The theory of the block universe sparks imagination. For instance, our dear ancestors could live in an eternal universe, forever separated from us.

18.5   Alien Life in the Multiverse

After exploring the philosophical, theological, and scientific origins of the Many Universe Hypothesis, one might ponder the relationship between alternate forms of life and the potential existence of an infinite array of universes. A simple inference suggests that if there are infinite universes, endless intelligent beings must also inhabit some of these universes—especially a universe whose laws are conducive to life. We may live in one of these universes where at least one intelligent species gets to appear: the human species. Moreover, there could exist universes where life is impossible due to the constraints of physical laws. This raises an intriguing question: What are the specific attributes of our universe that allow for life, and what do these characteristics imply about the likelihood of life existing in other universes?

18.5.1 Emergence of Life

Let us consider the work of Tomori Totani, a Japanese cosmologist. In a 2020 article, Totani tackles the problem of the emergence of life in an inflationary universe [279]. As we described in the second chapter, cosmological inflation is a theory that posits that the universe underwent an exponential expansion in its early stage due to the presence of a repulsive force. Totani tackles the problem of abiogenesis, a fundamentally crucial scientific mystery. How did life arise on our planet? Totani makes the initial assumption that the theory of the RNA world is valid. This theory is widely accepted and explains the appearance of DNA from an evolutionary standpoint. Then, he is left with the problem of how replicating RNA was formed randomly and abiogenetically. Replicating RNA requires a polymer of at least 40–100 nucleotides. His analysis leads him to conclude that such formation is possible in the entire universe, which includes the unobservable one. Totani shows us that an inflationary universe can have 10^{100} stars. From a probability standpoint, that universe is large enough to justify the formation of a replicating polymer. However, the problem is that the random formation of such large polymers is improbable in our observable universe with 10^{22} stars. The minimum number of nucleotides of a non-replicating RNA polymer in our smaller universe must be smaller than 20 nucleotides for life to have appeared on Earth abiogenetically with certainty. Therefore, there is no reason to expect more than one abiogenetic event in our observable universe.

Totani concludes that the discovery of extraterrestrial life could imply that other mechanisms, beyond randomness, are involved in the formation of replicating RNA. For instance, catalytic processes could speed up the formation of RNA molecules. Totani's theory bodes well for the existence of life in a multiverse. There could be alternative theories to Totani's assumptions.

Since life exists in the observable universe, it is possible that abiogenesis may not be the only mechanism by which life appeared on Earth 3.8 billion years ago. As mentioned before, panspermia may have played a critical synergetic role. While it does not ultimately explain the genesis of life, it could be an essential mechanism for seeding life in the universe. Furthermore, life in the universe may not necessarily be based on RNA/DNA or be carbon-based. Finally, life could be a phenomenon where the formation of complex molecules, from simple ones to ordered ones, may not be ruled solely by random factors.

18.6   Inflation and Multiverse

In a certain sense, we are sure that we live in a special multiverse, as we know an unobservable universe exists. According to cosmologist Alan Guth, the unobservable universe is incredibly large, perhaps infinite. Based on estimating its size before inflation, his calculations yield an unobservable universe 10^{23} times larger than our observable universe, an unfathomable size. Furthermore, there is a variation of the theory of inflation where the universe keeps expanding forever in what is called "eternal inflation" [280]. In fact, in 1983, American physicist Paul Steinhardt showed that, in some parts of the universe, the inflationary process does not end [281], a theory that he later opposed. The eternal inflation creates a series of bubble universes where the process of inflation decays, surrounded by other regions that keep inflating exponentially. We may live in one of these bubble universes. Since the space between these bubble universes is forever growing, no information from a bubble universe can exchanged with another. What are the physical laws of these bubble universe? No one knows; they may obey similar laws like our universe but with different values for the dimensionless constants of physics. Furthermore, the reader should understand that the eternal inflation multiverse is not connected to Everett's Many Worlds theory. These are two distinct and separate theories in two different fields. Steinhardt's multiverse is a much more physical concept than Everett's theory.

An inflationary universe can be seen as a collection of infinite observable universes depending on the observer, each separated from the others, a type of multiverse. Perhaps there are universes in the inflationary multiverse where the physical laws do not permit the existence of life. It is also possible that there is no physical speed limit in some of these universes that host life, which allows extraterrestrial civilizations to discover new worlds incessantly. But is it true that we live in a unique universe fit to host life? Here comes the Fine-Tuning argument (Fig. 18.1).

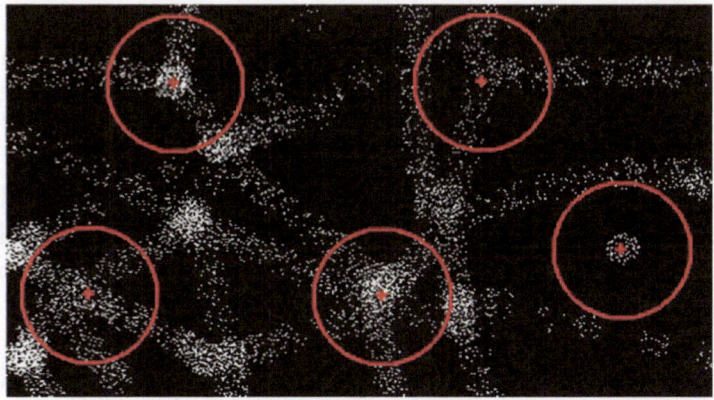

Fig. 18.1 Example of Multiverse. In the Universe, there are many observable areas or universes. The observable universes are marked as red-circled with a red cross in their center. By User: K1234567890y, Public Domain, https://commons.wikimedia.org/w/index.php?curid=1650827

18.7 The Fine-Tuning of the Universe

The argument of the Fine-Tuning of the Universe posits that the dimensionless constants fundamental to our understanding of the cosmos must possess values that allow for the existence of life. It suggests that even the slightest alterations to these constants would result in a universe where life could not emerge. This argument is frequently cited by believers and theologians as evidence for the existence of a creator, sparking debates among proponents and skeptics alike. American chemist Lawrence Joseph Henderson proposed a similar concept, though not identical, in his 1913 work "The Fitness of the Environment." In it, he explored how water's presence and unique properties, combined with Earth's chemical composition, are essential for sustaining life. His work is an example of a fine-tuning argument, as he pointed out that certain environmental conditions are necessary for life. While Henderson's argument aligns more closely with the Rare Earth hypothesis, it nonetheless raised awareness that the emergence of life on Earth might not be merely coincidental. The scientist who further developed the concept of the fine-tuning of the universe scientifically was Robert H. Dicke, a prominent American physicist of the latter half of the 20th century. Initially involved in radar technology at MIT's Radiation Laboratory, Dicke later transitioned to Princeton University, where he researched radiation and laser technology and tests for general relativity, such as the perihelion precession of Mercury [282].

Dicke observed that the expansion rate of the universe is finely calibrated. If the expansion were too rapid, the universe's density would become too low for stars to form. Conversely, if the expansion were too slow, the universe would eventually collapse in on itself in a "big crunch." In either scenario, life as we know it would not be possible. Therefore, the laws governing gravity, electromagnetism, the weak force, and the strong force must also be finely tuned for our universe to exist in its current state.

More recently, cosmologist and UK astronomer Royal Sir Martin John Reese authored a 1999 book, "Just Six Numbers: The Deep Forces That Shape the Universe," which focused on fine-tuning the universe [283]. In his book, Reese elaborates on six cosmological dimensionless numbers, emphasizing their finely tuned values that allow for the existence of life. He illustrates this by delving into the significance of a specific constant, which represents the ratio of the electromagnetic force to the gravitational force between two protons. We know that the electromagnetic force is vastly stronger than the force of gravity.

To illustrate why, let us consider that a single magnet can lift an object against the gravitational force generated by an entire planet. In this universe, the ratio between the electromagnetic and gravitational forces between two protons is unbelievable, denoted by ten followed by 36 zeroes. If this ratio were smaller, indicating stronger gravity, the universe would be much smaller and short-lived because stars would collapse more rapidly under their immense gravitational forces. Conversely, protons would repel each other too strongly if the ratio were larger to form heavier atoms. From a physicist's standpoint, the universe appears finely tuned. There are also valid arguments against the fine-tuning hypothesis. One argument is that these constants may exist because our understanding of the universe is incomplete and could vanish with the emergence of a comprehensive final theory encompassing all forces and particles. Another argument proposes that these dimensionless parameters might be correlated, meaning that modifying one parameter could necessitate changes to others or all of them.

18.8 An Alternative View of the Fine-Tuning Problem

As previously mentioned, our perception of the universe as finely tuned is based on our current limited understanding. It remains uncertain whether new, more comprehensive theories in physics will confirm or refute this notion. Additionally, humans have only recently begun to observe and contemplate this fine-tuning, a duration that is insignificant when compared to the billions of years that life has existed on our planet.

Was the appearance of *Homo sapiens* a foreseeable and necessary event since life appeared on Earth? Not at all, as evidenced by the numerous major extinction events that nearly wiped out all life on Earth. Therefore, any reasoning based on the survival of humans suffers from "survivorship bias," which occurs when only surviving samples are included in statistical reasoning. We cannot but observe a universe that permits life: the anthropic principle. This bias poses a challenge for the fine-tuning argument. In that case, where does this leave us? Considering the uncertain future survival of the human race due to internal and external cosmic threats, the anthropic principle may need to be extended to all intelligent forms of life in the universe. If intelligent life is abundant, then the issue of our potential finiteness in time becomes moot, as there would likely be other observers to replace us if we were to perish. Conversely, if intelligent life is rare, then the universe seems inefficient at generating observers to observe the nature of the universe. Alternatively, we could inhabit a universe that is not fine-tuned and does not require observers. In this case, we would happen to be the only intelligent species in this vast universe—an improbable albeit possible scenario. The anthropic principle is an extensive philosophical and cosmological subject with many contributors with different versions that we reserve for the final chapter. Another possibility is that we live in a subuniverse created by incredibly advanced extraterrestrial entities, leading us into the realm of the simulation hypothesis.

18.9   The Simulation Hypothesis

The so-called simulation hypothesis assumes that the world, the universe we observe, and everything in it is a "simulation" [284, 285].

This hypothesis might appear incredible and lacking in evidence, yet many scientists and philosophers hesitate to dismiss it outright. While leaving aside the specifics of what the term "simulation" might encompass, a similar concept has historical precedence in philosophy. For instance, René Descartes, a notable French mathematician and philosopher of the 17th century, put forth similar ideas. In 1641, he wrote "Meditations on First Philosophy," a book devoted to God's existence and the soul's immortality. In this book, Descartes reflects on how his senses have deceived him, leading him to doubt his ability to know things honestly. He systematically develops a method to eliminate doubt and attain certainty in his observations. Amidst his skepticism of the senses, Descartes entertains the notion of an evil demon capable of presenting him with an utterly illusory world. This powerful demon's objective is to deceive Descartes to the extent that he doubts everything his senses perceive, including his existence. Ultimately, Descartes reaches the famous conclusion "Cogito, ergo sum," or "I think, therefore I am," which serves as a foundation he believes cannot be shaken by any demonic trickery.

The Meditations embark on a journey of skepticism to establish a foundational understanding of knowledge free from doubt. Another philosophical concept, the "brain in a vat" hypothesis, has captured the imagination of many through its portrayal in works like the 1999 film "The Matrix." Coined by American philosopher Hilary Putnam [286], this hypothesis suggests that individuals are not embodied beings but brains suspended in vats, receiving simulated sensory inputs that create the illusion of a real world. This notion represents a form of global skepticism, casting doubt on the authenticity of our perceptions and experiences of the external world, suggesting that our understanding may be fundamentally flawed and deceptive. The simulation hypothesis, in light of advancements in computing technology, presents intriguing possibilities. As computational power grows, so does our ability to simulate natural phenomena with increasing accuracy. These simulations, spanning domains such as weather patterns, chemical reactions, and cosmological events, offer insights into the behavior of the physical universe.

The advent of neural networks has revolutionized our ability to predict and understand the intricate relationships within complex systems. Simultaneously, there's a growing trend, particularly among younger generations, to be constantly connected to digital devices like phones or laptops, engaging in activities ranging from communication to immersive virtual experiences, such as the game Minecraft. In Minecraft, players enter a meticulously crafted 3D virtual world where they can interact with other players and virtual entities for entertainment. A feature of these virtual worlds is that they are discretized in space and time since a computer can't simulate a continuum. This discretization bears an analogy to the natural world, where a similar albeit much finer discretization occurs, as described by the principles of quantum mechanics with the existence of particles and quanta. As computational power continues to advance exponentially, it is conceivable that humanity will create increasingly sophisticated virtual environments populated by simulated beings, allowing researchers to study their behavior in controlled settings within these virtual words.

If the simulation hypothesis is correct, it is also possible that virtual beings live in a mega universe built like a *matryoshka*, a Russian doll, where one alien species creates a simulated universe whose inhabitants also decide to make their own simulated universe and corresponding inhabitants and so on. As we go deeper into these nested simulations within the mega universe, a notable trend emerges: with each successive layer, the fidelity of the simulation diminishes, resulting in increased blurriness and less finely discretized representations of reality. This phenomenon arises from the inherent limitation that a universe within another can never possess a finer level of discretization than its containing universe.

Moreover, as alien civilizations advance to higher levels within the nested simulations, they become increasingly adept at minimizing bugs and software errors within their simulations. Consequently, inhabitants of a given simulated universe could potentially gauge their position within the matryoshka mega universe by examining the discreteness and accuracy of their universe's laws. The less bug-free universe is higher on the scale of the nested universes.

18.10   A Strange Universe

Exploring various intriguing hypotheses about reality, including multiple universes where extraterrestrial life could potentially thrive, is captivating. While it is widely acknowledged that advanced alien species have not visited Earth, let us entertain the notion of a reality where such encounters have occurred or might still be possible.

This is the subject of the next chapter.

19

A Daring Hypothesis: Alien Visitation

This chapter departs from the others by crossing into science fiction rather than grounded reality. It aims to spark imaginative scenarios for our readers and modern writers rather than present factual accounts of the existence of extraterrestrial life. Thus far, no conclusive evidence has supported the idea that Earth has been visited or is currently being visited by extraterrestrial beings. While there is anecdotal evidence from individuals who claim to have witnessed strange phenomena in the sky, such as unidentified flying objects, there is no definitive proof, only ambiguous footage in some cases. Additionally, there have been claims of abduction by aliens, often portrayed in a negative light. However, this book does not address these stories without scientific proof.

Despite the lack of concrete evidence, readers are encouraged to think creatively and consider the possibility of alien visitation in the past or even more recently. This out-of-the-box thinking is what scientists do to cover all possible hypotheses.

19.1 Microorganism Visitation

The absence of radio communication from intelligent species from another planet is not equivalent to saying that we are alone or have never been visited.

When considering alien life, it is instinctive to conjure up images of sci-fi scenarios involving formidable spaceships traversing the cosmos, perhaps carrying alien armies. However, alien life could encompass something simpler, like viruses and bacteria. These microorganisms might be based on a combination of RNA and DNA. They could also be based on different chemistry and biol-

© The Author(s), under exclusive license to Springer Nature Switzerland AG 2025
L. Vacca, *Life Beyond Earth*,
https://doi.org/10.1007/978-3-031-81695-6_19

ogy. They could lie dormant in the depths of our oceans. Indeed, there could be all kinds of primitive forms of life developing on exoplanets and exomoons outside our solar system. NASA has considered the possible presence of simple alien forms on our moon. Early Apollo mission astronauts have spent time in quarantine in case of possible contamination with moon substances and alien life, a practice that has been eventually discontinued.

It is quite conceivable that some of these primitive organisms could cross the solar system inside pieces of rocks and land on Earth.

When a large meteorite or asteroid impacts a planetary object like Earth, it can eject planetary material into orbit. This ejected material may travel to other planets or moons within the Solar system, or it could be propelled out of the system into interstellar space by the gravitational forces of gas giants like Jupiter. Within this material, extremophiles—tiny organisms resistant to harsh space conditions—may be present and could survive, at least partially, the impact from another celestial body. We have also seen that tardigrades can survive space travel thanks to their ability to control their metabolism. Therefore, Earth has been dispersing organic material into the cosmos in its simplest form and receiving similar material in return, perhaps from the exomoons of Jupiter, Saturn, and Mars in the distant past.

Organic material may have originated from other solar systems and previously traveled across interstellar space carried by rocks not bound by solar gravity.

Some of this material might have been DNA-based and fully compatible with existing forms of life, while other, more unconventional forms of life may not have thrived in Earth's conditions, leading to their degradation and demise. Until now, all known forms of life humankind encounters are based on RNA and DNA. However, speculation about alternative forms of life remains just that—speculation. There is no evidence that there are alien forms of life living on our planet.

What about intelligent species? Is it possible that extraterrestrial beings have visited us in the past? Why would they go undetected?

19.2 A Thought Experiment

Many of us as kids have had lots of fun watching how ants operate. There is no doubt that ants are a clever species. These formidable tiny insects form colonies with thousands, if not millions, of single ants, each busy carrying out their task. Whether searching for food, protecting the colony from other predatory insects, or reproducing themselves, ants are present virtually in every corner of the Earth except for Antarctica, polar and subpolar environments. And yet, when we walk in a garden or a forest, we often do not notice or directly interact with ants unless we encounter their colonies or experience their stings. This suggests that there may be deeper reasons underlying our infrequent interactions with ants.

One could be that ants are fundamentally beings that live only on a two-dimensional surface or underneath the surface, while we live in a three-dimensional world. Another reason could be their size. Many life forms surround us daily, yet we are often unaware of their presence. Plenty of microorganisms inhabit our bodies' skin and interior organs despite us rarely sensing their existence.

This begs the question: would we be capable of recognizing the presence of alien life and technology without expecting it? In some cases, the answer is yes, mainly if an alien being displays technology that dramatically interacts with the environment. For instance, technology could create pockets that display complex order or negative entropy. However, there could be some instances where the alien possesses a technological level far superior to us and would not want to be discovered. Consider, for a moment, the possibility that our universe might differ significantly from our perceptual understanding. Let us explore the perspectives of thinkers who have challenged our conception of reality.

19.2.1   Do We Live in a Multidimensional Universe?

Throughout history, humanity has harbored a suspicion that there may be more to the world than what our senses reveal. Consider an example from philosophy: Plato, a renowned Greek philosopher who lived around 400 years before Christ, left behind a significant body of written work. His most famous piece, "Republic," penned around 375 BC, is based on dialogues between Socrates, Plato's teacher, and other Athenians, discussing justice, political systems, society, and various philosophical topics. Socrates, known for his Socratic method of questioning to test beliefs and knowledge, features prominently in "Republic." In Book VII of "Republic," we encounter the famous Allegory of the Cave. Plato writes about Socrates teaching Glaucon, one of his disciples. In one scene, Socrates describes a scenario of prisoners confined in a dark cave facing a wall their entire lives since birth. Due to their restricted positioning, these prisoners can only see shadows projected on the wall before them. Believing these shadows to be reality, the only reality, the prisoners remain unaware that they are mere projections of objects, puppets, and a fire behind them. The Allegory of the Cave provides various metaphysical, epistemological, sociological, and political interpretations. One possible inference from this allegory is that beyond the realm of our senses lies a reality entirely different from our experiences. This concept is not unfamiliar to religions either. Let us look at the example of Buddhism and some schools of Hinduism. The idea of *Maya* considers the material world a veiled illusion. The world we see, hear, and touch appears real and solid. This world is what appears through our senses. However, according to Maya, what we see is only a veil of the actual reality that is constantly changing.

Even in physics, numerous scientific theories posit the possibility of dimensions beyond the four dimensions of Einstein's spacetime continuum to elucidate reality.

For instance, enter German mathematician and physicist Theodore Kaluza. In 1919, Kaluza merged Einstein's theory of gravitation with electromagnetism, utilizing a five-dimensional space [287]. He shared the outcome with Einstein, who encouraged him to publish it, leading to its publication in 1921. Subsequently, in 1926, Swedish physicist Oskar Klein further developed the theory, suggesting that the fifth dimension forms a circle of nearly infinitesimal size and exploring its potential quantum interpretation [288]. The Kaluza–Klein theory is among several unification theories that incorporate extra-dimensional spaces. Moreover, consider the more contemporary instance of superstring theory. This theory seeks to unify forces by treating particles as vibration modes of minuscule strings. The theory operates within a 10-dimensional space to maintain mathematical consistency, wherein six dimensions are compactified to sizes smaller than an atom [289]. In contrast, the remaining dimensions resemble the conventional spacetime familiar to us. Furthermore, we can find an M-theory, named for "membranes" or "branes," which posits an 11-dimensional spacetime where reality may entail additional finite dimensions akin to those we perceive [290] (Fig. 19.1).

Do these dimensions exist, or is it merely the consequence of mathematical elegance?

In contemplating such profound inquiries, some physicists caution against placing excessive faith in pursuing mathematical elegance. In contrast, others posit that our reality transcends a 3D space and a singular time dimension. For the sake of such a hypothesis, let us imagine that the latter perspective holds and that concealed dimensions exist beyond our perceptible realm. For simplicity's sake, let us envision the existence of an additional spatial dimension alongside the three familiar dimensions. In such a scenario, any object residing within this 4D space may or may not intersect with our 3D space, allowing objects to materialize and vanish within our perceptible realm. To illustrate this concept, consider an ant traversing a flat pavement, existing and interacting

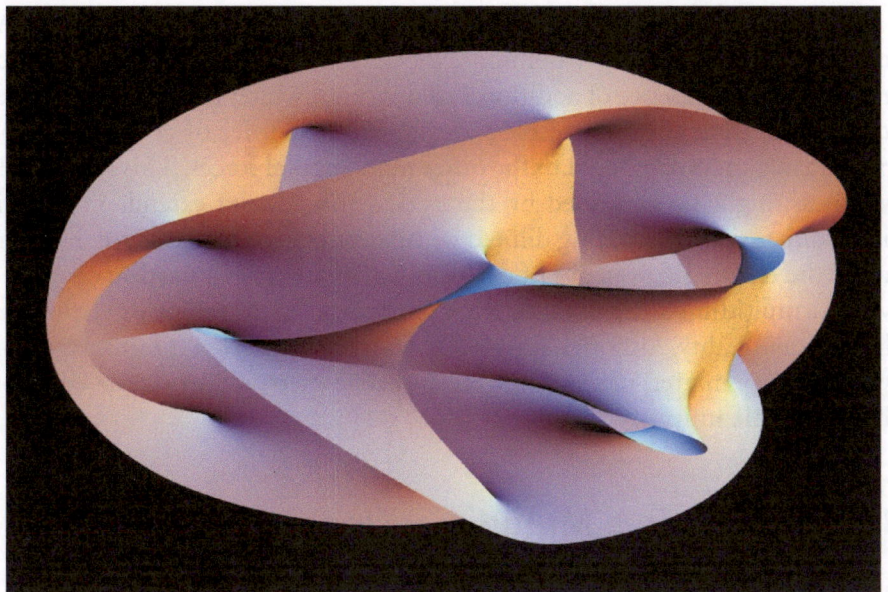

Fig. 19.1 Artistic rendition of an eleven dimension spacetime. By TexasVenom, CC BY 4.0, https://commons.wikimedia.org/w/index.php?curid=131714029

solely within the plane of the pavement. Unless we intrude upon its plane, the ant remains oblivious to our presence. At the same time, we, in turn, can observe it, provided that light from its dimension can penetrate ours.

Moreover, as we start walking on the pavement, the ant would perceive the outlines of our feet materializing instantaneously while the remainder of our body remains unseen. From the ant's perspective, the space on the pavement appears to be occupied by a solid object as our feet make contact. However, as we elevate our feet to climb the stairs, the ant witnesses the sudden disappearance of the two foot-shaped objects. Thus, for beings confined to perceiving subdimensional space, conventional laws governing the conservation of matter and energy may no longer hold. Nevertheless, these transient appearances and disappearances might occur at such rapid intervals that, on average, the conservation principles may remain applicable. Within theoretical physics, we encounter the concept of "virtual particles," entities theorized to emerge and vanish randomly within limited spatial and temporal constraints, resulting from the indeterminacy principle in quantum mechanics. Sometimes, these virtual particles can interact with other particles before disappearing. The existence of virtual particles is at the basis of the existence of vacuum energy, an energy found even in the deepest intergalactic vacuum that could be responsible for the universe's expansion.

Do these virtual particles exist, or are they just mathematical artifacts? As usual, there are discordant opinions. However, their hypothetical existence may justify the presence of additional dimensions, which is fascinating. Other mysteries hinting at unseen dimensions include the force of gravity's striking extreme weakness compared to the other fundamental forces. This remarkable weakness might be explained by the notion that gravity extends into additional dimensions, becoming diluted within our observable universe. Furthermore, the enigmatic presence of dark matter, behaving as a particle concealed within a multidimensional realm, offers another potential indicator of a higher-dimensional reality. Although this elusive entity refrains from direct interaction with ordinary matter, its gravitational influence exerts a measurable effect on the observable universe.

19.3   Beings in a Multidimensional Universe

We have listed a few possible conjectures that support the existence of one or more dimensions we cannot perceive. Let us now consider the existence of intelligent beings dwelling within a multidimensional realm, such as a 5D continuum—one dimension beyond our own. For instance, imagine humans in our 4D space and those occupying the 5D space, moving about randomly. In such a scenario, their paths would likely never intersect, as the volume of our 4D space, perceived from their perspective, would be insignificant. To illustrate further, consider a person drawing a line across a vast 2D universe. Along this line reside unique beings constrained to exist solely within this 1D universe. The likelihood of any ant traversing the 2D space at random, encountering and remaining on this line long enough to encounter one of these singular beings is also tiny.

Therefore, meetings between beings living and interacting in a space with different dimensions can only be intentional and initiated by beings living in a space with more dimensions. Even when this type of encounter becomes possible, as just described, the special beings in 1D will never really be able to see the real appearance of the 2D beings, only the projection of the higher dimensional beings onto their world.

The strangeness of these hypothetical multi-dimensional encounters is exacerbated by the potential for higher-dimensional beings to manipulate reality during a "visit" to lower-dimensional spaces and then vanish without a trace once their task is completed. These modifications might transpire without the direct presence of the alien entity, instead changing fields emanating from their section of reality that borders ours. Another intriguing possibility arises when two objects, separated by great distances in their world, are intricately linked in a higher-dimensional space, such that each affects the other and vice versa. This concept is reminiscent of curled-up spaces, where objects could be closer in a higher-dimensional context than in a lower-dimensional one. This scenario resembles the phenomenon of quantum entanglement and its "spooky action at a distance," which was previously discussed in terms of communication among civilizations. Furthermore, the laws of physics governing a higher-dimensional world may differ entirely from those in our own. As previously noted, gravity may exert a more powerful force in the extra-dimensional space inhabited by aliens, and lifeforms may exhibit entirely different chemical compositions and behaviors.

19.4   Searching for Answers

We have explored some interesting notions that offer alternative views of our universe, challenging the limitations imposed by our senses. This exploration is motivated by the absence of scientific evidence definitively proving that we are not alone. Therefore, to address one of the most fundamental scientific and philosophical inquiries, we must entertain all possibilities, even the most fantastical ones.

We conclude this chapter repeating the immortal words written by English playwright Shakespeare and spoken by Hamlet to his friend Horatio: "There are more things in Heaven and Earth, Horatio, than are dreamt of in your philosophy."

20

The Case for Extraterrestrial Life

The Fermi paradox, a captivating enigma that has intrigued philosophers, sci-entists, and the public, presents a variety of potential explanations. The ones discussed in this book were chosen for their prominence in the ongoing and ever-evolving discourse. It's crucial to understand that, at present, a definitive answer to the Fermi paradox remains elusive. The question of the existence of other intelligent beings in distant realms carries profound implications, stretching beyond the boundaries of science to encompass profound meta-physical considerations.

What kind of universe do we live in? What is the role of the human race in the universe?

Are we the product of an incredibly rare occurrence, or are there countless corners of the universe where alien life forms are evolving? Human existence is a testament that the laws of the universe allow for the emergence of intelligent life. It's within the realm of possibility that these same laws could nurture intelligent life forms vastly different from ours. As the famous astrobiologist Carl Sagan suggested, the universe might seem wasteful and improbable if we were the sole intelligent species in the universe. However, despite its improbability, this possibility cannot be dismissed outright. Given the fantastic odds of being the only civilization in the universe, one may conclude that we live in a sort of experiment, as the simulation hypothesis suggests.

However, it is much more probable that advanced civilizations perish quickly when they encounter the Great Filter. Hence, they do not last long enough to reach the technological level where they can send powerful signals throughout our galaxy or build Dyson spheres. Recently, the notion that we are not alone has gradually gained traction worldwide. Indeed, humanity's understanding of

L. Vacca, *Life Beyond Earth*,
https://doi.org/10.1007/978-3-031-81695-6_20

the universe underwent a profound shift with the introduction of the Copernican system, suppressing the geocentric models of Aristotle and Ptolemy. Let us take a quick look at the life of this great astronomer.

20.1   The Life of Nicolaus Copernicus

Nicolaus Copernicus, born in Torun, Poland, in 1473, was a notable Polish mathematician and astronomer who proposed the heliocentric model of the universe, which placed the Sun rather than the Earth at its center [291]. Copernicus lost his father, a wealthy merchant, at ten. Copernicus was taken in by his maternal uncle, Lucas Watzenrode, a prominent cleric who provided him with support and guidance. In 1491, Copernicus pursued his education at the University of Cracow, where he delved into mathematics and astronomy. During his university studies, Copernicus learned about Ptolemy's astronomic system. In 1495, Copernicus was given a position of Catholic canon by his uncle, a Prince-Bishop, and a significant administrative responsibility within the Church. However, due to internal solid opposition to his appointment, Copernicus moved to Italy to study as part of his educational formation. Copernicus traveled to Bologna, where he furthered his studies at the renowned University of Bologna under the tutelage of Italian astronomer Domenico Maria Novara da Ferrara. During his time in Bologna, Copernicus conducted numerous astronomical observations and immersed himself in Greek, enabling him to explore ancient Greek astronomers' works. While in Bologna, Copernicus encountered discrepancies within the prevailing Aristotelian and Ptolemaic cosmological frameworks. His dissatisfaction with these systems prompted him to immerse himself in astronomical theories. In 1500, Copernicus relocated to Rome, where he delivered mathematical lectures on astronomy. He later pursued a medical degree in Padua (Padova) from 1501 to 1503. Returning to Warmia in 1503, Copernicus held various roles, including physician, administrator, and economic advisor, all while continuing his work on developing a heliocentric model of the universe throughout his life. He put forward the basic idea behind the economic theory known as "Gresham's law." Such theory states that bad money chases out good money in a two-currency economy. His passion for astronomy led him to conduct many observations of the planets in our solar system, which were instrumental in developing his heliocentric system. He started working on his heliocentric system while residing at his uncle's castle at Lidzbark Warminski, and around 1509, completed a short description of his system, the "Commentariolus." In 1533, a secretary to the Pope, Johann Widmanstetter, taught Copernican's heliocentric views to Pope Clement VII and his curia. His views encountered a strong response from the Archbishop of Capua, who advised Copernicus to renounce his theories. Copernicus started writing his main work, "De revolutionibus orbium coelestium," in 1517 and worked on it throughout his life until his death. "De revolutionibus orbium coelestium" was completed and published in 1543, although Copernicus feared publishing it due to his fear of religious backlash.

A legend says that Copernicus, waking up from a coma on his deathbed, died right after seeing a printed copy of his "De revolutionibus orbium coelestium." Nicolaus Copernicus is buried in the Frombork Cathedral in Poland.

Copernicus' work ignited the Copernican Revolution, marking a significant turning point in European scientific thought and ushering in a new era of understanding of the cosmos (Fig. 20.1).

20.2 The Copernican Principle

The Copernican principle, stemming from Copernicus's heliocentric model, posits that nothing is inherently unique about our position within the universe. American astrobiologist Carl Sagan noted that our solar system resides in an unremarkable location within the Milky Way galaxy, which is not distinguished among galaxies in size, shape, or cosmological significance. It is worth quoting the words of this great scientist: "We live on an insignificant planet, of a humdrum star, lost in a galaxy, tucked away in some forgotten corner of a universe." Furthermore, the laws of physics, including those of special relativity, adhere to the Copernican principle by asserting the absence of any privileged frame of reference or observer. We have seen in the second chapter that the universe is probably homogeneous and isotropic at vast scales by observation; as a consequence of this principle, modern cosmology models start from such assumptions.

Experimental physics requires that the result of an experiment should be reproducible regardless of where and when unless there is a specific reason for it. The principle of reproducibility states that an experiment conducted on Earth will yield the same results anywhere in the universe. Such a principle is the basis of the Copernican view.

Giordano Bruno, a Catholic Italian friar, philosopher, and cosmologist who lived in the 16th century, used the central ideas of the Copernican system to go even further.

He proposed that the universe was infinite, populated by countless suns and planets, many of which might harbor life similar to Earth. Bruno's insight also led him to assert that the universe lacked a central point. However, the

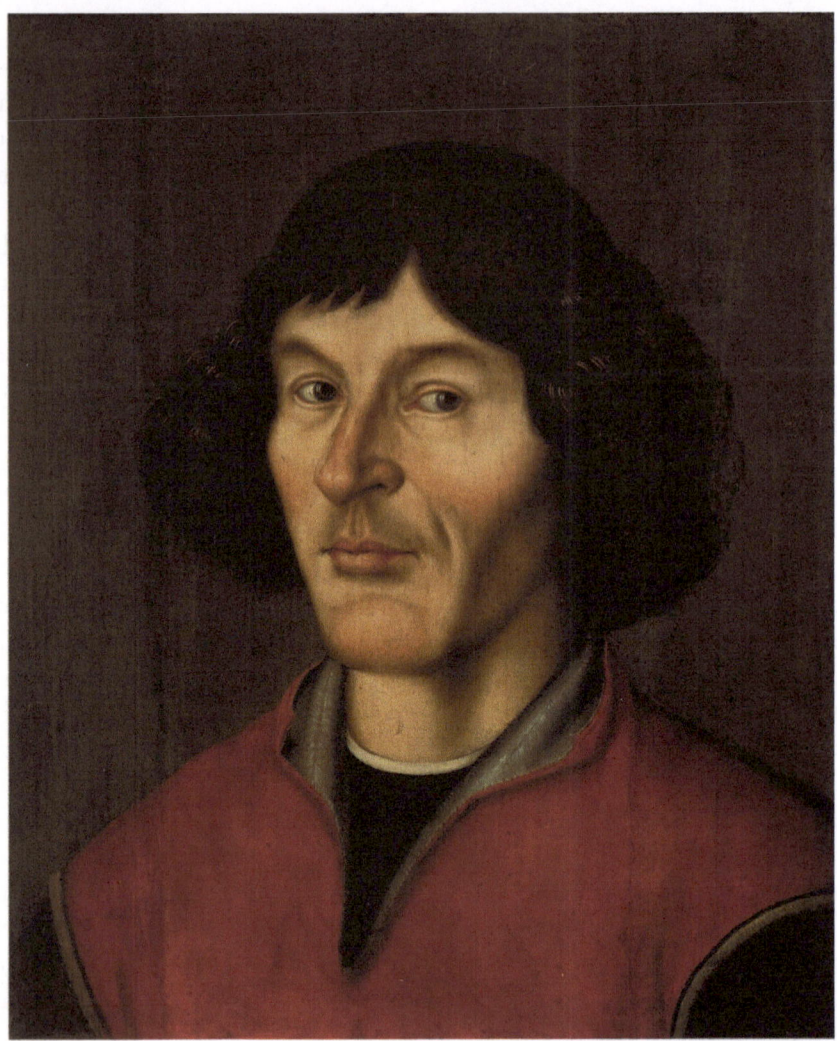

Fig. 20.1 Nicolaus Copernicus portrait from Town Hall in Torun—1580. By an unknown author.—http://www.frombork.art.pl/Ang10.htmhttps://www.welt.de/img/kultur/mobile152954235/1212503297-ci102l-w1024/Kopernikus-Gemaelde-in-Krakau.jpg, https://muzeum.torun.pl/wp-content/uploads/2023/03/Portert-Mikolaja-Kopernika-MOT.jpg,   Public   Domain,   https://commons.wikimedia.org/w/index.php?curid=147956893

Fig. 20.2 Giordano Bruno. Public Domain, https://commons.wikimedia.org/w/index.php?curid=52557

Catholic Church's Inquisition met his views with hostility, which sought to suppress teachings deemed heretical to the faith. Bruno faced trial and was ultimately condemned for heresy. He met a tragic end, being burned alive at the stake in Campo de Fiori, Rome. Campo de' Fiori visitors can see a statue honoring Giordano Bruno's memory today (Fig. 20.2).

Let us now consider the fractality of nature as a counterargument to the Copernican principle.

20.2.1   Fractals in Nature

One typical counterargument to the Copernican principle suggests that the universe lacks perfect isotropy, meaning uniformity in all directions. It contends that observable variations and structures are present in different regions of the cosmos, even on a cosmic scale. Additionally, certain recurring patterns are observed across various scales within the universe. For instance, in nature, we often observe *fractals*, a term coined by French mathematician Benoit Mandelbrot, who made significant contributions to their understanding [292].

A fractal possesses a unique property known as "self-similarity," meaning its patterns repeat across all scales. For instance, when examining a snowflake under a microscope, one would notice patterns resembling larger-scale snowflakes. Nature offers numerous examples of fractals, including snowflakes, coastlines, crystals, and DNA. Fractals can emerge from both deterministic equations and random processes like Brownian motion. While there is ongoing theoretical debate about whether the universe functions as a colossal fractal, empirical observations indicate that the distribution of galaxies at a scale of 50 million light-years displays fractal traits [293]. It is important to note that while the fractal scale is quite large, it is smaller than the scale at which astronomers deem the universe homogeneous and isotropic. Furthermore, as we have seen, the universe is not static. Instead, it expands in all directions while maintaining its homogeneity and isotropic characteristics.

20.3   The Anthropic Principle

As intelligent life resides on Earth, our planet's solar system possesses a distinct set of characteristics that, at the very least, appear unique compared to the limited sample of discovered exoplanets. This is consistent with the anthropic principle previously discussed in this book. The apparent uniqueness of our solar system seems to contradict the Copernican principle. However, we can endeavor to find common ground between these seemingly contradictory notions. An alternative perspective of the Copernican principle proposes that despite Earth and our solar system possessing distinct traits, they should not be solitary entities in the universe. Instead, they should be accompanied by numerous solar systems and planets with similar, albeit not identical, features conducive to life. Under this interpretation, Earth and the universe are no longer mere run-of-the-mill entities solely because we inhabit them. Perhaps a potential resolution lies in the uniqueness of Earth and the universal constants, which, while singular, do not preclude the existence of "similar" worlds or universes.

An argument preceding the anthropic principle was initially proposed by the physicist mentioned above, Robert H. Dicke, who also made remarks regarding the fine-tuning of the universe, possibly to challenge the implications of his observations [294].

The term "anthropic principle" was coined by Australian astrophysicist Brandon Carter, who extensively researched this concept [295]. Carter thinks our situation is somewhat fortunate but not incredibly unusual. Carter coins this argument as "a weak form of the anthropic principle." The weak form advocates an intermediate view between the cosmological principle and geocentrism, where observers can be other intelligent civilizations. The privilege we enjoy is our position in space and time. The weak form also suggests that the principle applies only during specific times and locations in the universe, mainly when a star goes through its main sequence. This weak form is somewhat consistent with the *Copernical principle*, which states that there is no preferred place or time in the universe and does not bar the existence of a solar system like ours.

The anthropic principle also has a strong form. Carter's strong form states that the physical parameters integral to the universe's laws have to permit the existence of intelligent observers at some point during the universe's history. He also introduces the idea of a multiverse with universes whose fundamental parameters are biofriendly.

One exciting version of the strong form is the work of English cosmologist John Barrow and American cosmologist Frank Tipler. In their 1986 book, "The Anthropic Cosmological Principle," the authors argued that a single universe must be uniquely structured to support life at some point, thereby facilitating the existence of conscious observers who can observe the universe. In these analogous realms, alternative life forms may thrive, differing from us, yet they are "alive and conscious." Their argument favors a version of the weak anthropic principle supporting a theological explanation for the numerous coincidences that the physical laws present us.

In addition to the various forms of the anthropic principle, what compelling argument supports the presence of life in the universe?

20.4 Adaption

Life is highly resilient and adaptive. However, Earth holds a special status primarily because we have yet to discover other worlds supporting life. As previously discussed, extremophiles exemplify life's adaptability, thriving in environments with extreme temperatures and pressures. Hence, a planet similar to Earth but with slightly different conditions, such as temperature or chemical composition, could harbor life. Another striking illustration of life's resilience arises from research conducted near Chernobyl (Chornobyl), the site of a significant nuclear power plant disaster. Despite the radiation levels rendering a 1000 square mile zone, known as the Chornobyl exclusion zone, uninhabitable for humans, scientists have uncovered life persisting in this harsh environment. Scientists, led by Sophia Tintori of NYU, have been collecting samples of the worm species *Oscheius tipulae* that live in the soil [296]. The unforeseen and astonishing discovery was that the worms' genomes remained largely intact despite the heightened radiation levels in the area. There could be a biological repairing mechanism that allows such worms and other species to survive in one of the most dangerous environments on Earth. This underscores life's extraordinary capacity to acclimate to even extreme environments. A thorough understanding of this adaptation mechanism could be harnessed to shield astronauts from radiation exposure during their future extended space voyages. Another significant mystery is that life forms may not necessarily rely on carbon as their fundamental building block. Some scientists have speculated that silicon, rather than carbon, could be an alternative basis for life.

20.5 Silicon and Life

Carbon is a foundational element capable of forming extensive chains, providing the essential structure for biological compounds. This versatility is evident in carbon's ability to create robust, enduring structures, exemplified by the hardest natural carbon-based substance, diamond. Conversely, heteroatoms like oxygen contribute to chemical diversity and directional bonding. Silicon, similarly, can act as a scaffolding element and is the second most abundant element on Earth after oxygen. It shares similarities with carbon, possessing a valence of four and the capacity to bond with four other atoms to form structures resembling carbon ones. However, the comparison diverges in the realm of biochemistry. While carbon combines with oxygen to produce carbon dioxide, a gaseous compound easily eliminated as waste, silicon reacts with oxygen to form solid, inert silicate compounds found in rocks and sand. Silicon's characteristics, including its crystallization at high temperatures and inability to form chiral compounds essential for life, set it apart from carbon. Nonetheless, silicon may still play a significant role in the biochemistry of alien life. In carbon-based life forms, silicon aids in various biological functions, such as plant reproduction and human bone health, suggesting its potential importance in alternative biochemical processes. It has to be said that we have yet to encounter silicon-based life. For the reasons mentioned above, the quest for life primarily focuses on detecting the foundational elements necessary for carbon-based life.

20.6 Searching for Life in the Solar System

Saturn's sixth largest moon, Enceladus, harbors a significant global subsurface ocean beneath a thick icy layer, with a concentration around its south pole region. This moon exhibits volcanic activity driven by internal heat from gravitational tidal forces. Ice volcanoes (cryovolcanoes) on Enceladus eject a mixture of gases into space, including water vapor, hydrogen, ice, and crystals. From 1997 to 2017, NASA's Cassini spacecraft conducted multiple flybys of Saturn and its moons, including Enceladus, during its mission. Cassini's observations revealed evidence of hydrothermal activity within Enceladus' gas plumes. In 2023, an international team of scientists led by planetary scientist Frank Postberg analyzed data from Cassini's Cosmic Dust Analyzer, specifically focusing on the mass spectra of ice grains present in the plumes [297]. Their analysis detected the presence of phosphorus in the form of sodium phosphate compound within the ice grains. These findings were subsequently published in Nature in 2023 under "Detection of Phosphates Originating From Enceladus' Ocean." It was assumed that phosphorus was absent from the moons orbiting gas giants. However, this discovery of phosphorus on Enceladus suggests a higher probability that other moons may also contain this essential element. This constitutes a significant discovery in the field of exobiology, which will lead to further exploration of this exciting moon. There is no conclusive evidence that life exists on these moons, but these results are encouraging. Furthermore, such finding increases the likelihood of finding microbial life within Enceladus's hydrothermal vents. Adding to the mystery of what the moons of gas giants may hold, Europa, one of Jupiter's moons, presents intriguing characteristics [298]. Europa, slightly smaller than Earth's moon, possesses a thin atmosphere and features a remarkably smooth surface, distinct from many other moons known for their volcanic activity. Numerous ridges mark its surface, comprising an icy shell that overlooks a salty ocean and likely contains more water than all of Earth's oceans combined. The presence of an internally generated magnetic field [300, 301] is likely a result of electrically conductive water within Europa induced by varying Jovian magnetic fields as Europa orbits Jupiter. With water and essential elements such as hydrogen, oxygen, carbon, nitrogen, phosphorus, and sulfur, Europa provides all the necessary ingredients for life [299]. Considering these factors, it is reasonable to speculate that many moons in our galaxy harbor water and have the potential to support life. We speculate that moons in our galaxy may contain water like some of the Jovian moons and that some may host life. While we have confirmed the existence of many exoplanets, the search for exomoons is very challenging. How long will it take before we find evidence of extraterrestrial life?

20.7   A Great Enigma

Space is mostly empty. The vast emptiness of space plays a crucial role in sustaining life as we know it. In a universe teeming with more matter than in our universe, frequent collisions would ensue, causing widespread chaos and eventually leading to the universe's collapse. Moreover, the speed of light, albeit breakneck, pales compared to the vast expanse of the universe and the relatively brief timespan during which intelligent life has emerged on Earth. To put this into perspective, the discovery of fire dates back at least a million years ago [302]. This inherent limitation on travel speed poses significant challenges to the likelihood of physical encounters with extraterrestrial beings. The immense distances and impediments associated with space travel make such encounters highly improbable. Furthermore, the existence of intelligent civilizations may be imperiled by potential existential threats, known as the Great Filter, which could lead to their extinction before contact with other civilizations is possible. Even in the event of receiving a radio signal from an intelligent species and embarking on a journey to meet them, they may have already gone by the time we arrive. Therefore, the presence of intelligent life in our galaxy is likely exceedingly rare.

Conversely, microbial life is expected to be far more abundant in the galaxy, existing in vast numbers. There are one trillion microorganism species or more for one intelligent, technologically advanced animal species on Earth [303]. Based on a conservative estimate derived from the Drake equation, which predicts the likelihood of extraterrestrial civilizations, there may be around ten other intelligent civilizations in the galaxy. Extrapolating from these statistics, there could be approximately ten trillion species traversing the galaxy through a process known as panspermia. Consequently, if intelligent life is scarce in our galaxy, our prospects of discovering microbial life on distant moons are considerably higher than receiving signals from an advanced civilization.

The reader should know that the search for extraterrestrial intelligence began relatively recently, in the 1960s, spurred by the space race between the USA and the USSR. Since then, numerous space probes have been launched to explore our solar system, yet there is still much left to discover in this realm of exploration. Predicting when we might uncover evidence of extraterrestrial life remains uncertain. While we know that there are numerous solar systems in our galaxy with planets and moons similar to those in our solar system, the process by which life emerges from inorganic materials and the likelihood of alternative forms of life on other celestial bodies remain unknown. If we were to venture a guess, one approach could be the doomsday argument mentioned before. Given that we have been actively searching for extraterrestrial life for the past six decades, starting with the first space missions, there is a 95% confidence interval that we may find approximately within the next 1.5–2,400 years. However, such estimates should be taken cautiously, as they rely on a heuristic probabilistic model. No one can be certain how long, if ever, it will take us to find extraterrestrial life.

In conclusion, we have seen that the essential elements of life are and will be found in innumerable exoplanets and, perhaps, many exomoons. We have shown that life is resilient and can thrive in challenging environments. This bodes well for the existence of microorganic and intelligent life in the universe. We may be too primitive for extraterrestrial civilizations that are millions, if not billions, years older than us. If such civilizations exist, their sole presence leads me to think that magnificent scientific and technological discoveries await us in the future.

Bibliography

1. https://www.sif.it/riviste/sif/sag/recensioni/lanouette
2. https://web.uniroma1.it/museofisica/enricofermiscuoladiroma
3. http://hyperphysics.phy-astr.gsu.edu/hbase/quantum/fermi2.html
4. https://www.atomicarchive.com/history/manhattan-project/p1s4.html
5. https://sgp.fas.org/othergov/doe/lanl/la-10311-ms.pdf
6. https://science.nasa.gov/missions/hubble/hubble-reveals-observable-universe-contains-10-times-more-galaxies-than-previously-thought/
7. https://arxiv.org/pdf/2011.03052
8. https://asd.gsfc.nasa.gov/blueshift/index.php/2015/07/22/how-many-stars-in-the-milky-way/
9. https://www.esa.int/Science_Exploration/Space_Science/Herschel/How_many_stars_are_there_in_the_Universe
10. Louis Strigari et al. ia, Nomads of the Galaxy, Monthly Notices of the Royal Astronomical Society, Volume 423, Issue 2, Pages 1856-1865. 2012
11. https://exoplanetarchive.ipac.caltech.edu/docs/counts_detail.html
12. Cassan, A.; Kubas, D.; Beaulieu, J. P.; Dominik, M; et al. (2012). "One or more bound planets per Milky Way star from microlensing observations." Nature. 481 (7380): 167-169. arXiv:1202.0903
13. Kipping, D., Bryson, S., Burke, C. et al. An exomoon survey of 70 cool giant exoplanets and the new candidate Kepler-1708 b-i. Nat Astron 6, 367-380 (2022).
14. https://science.nasa.gov/solar-system/moons/
15. https://science.nasa.gov/moon/formation/
16. arXiv:1605.07178v2 [astro-ph.CO]

17. Javanmardi, B.; Porciani, C.; Kroupa, P.; Pflamm-Altenburg, J. (August 27, 2015). "Probing the Isotropy of Cosmic Acceleration Traced By Type Ia Supernovae". The Astrophysical Journal Letters. 810

18. arXiv:2206.05624 [astro-ph.CO]

19. arXiv:2207.05765 [astro-ph.CO]

20. Slipher (1913): Lowell Observatory Bulletin, 58, 56.

21. Slipher (1914): Lowell Observatory Bulletin, 62.

22. Slipher (1915): Popular Astronomy, 23, 21.

23. Slipher (1917): Proc. Amer. Phil. Soc., 56, 403.

24. Edwin Hubble, "A relation between distance and radial velocity among extragalactic nebulae," From the Proceedings of the National Academy of Sciences, Volume 15, March 15, 1929, Number 3.

25. Albert Einstein : "Zur Elektrodynamik bewegter Körper", Annalen der Physik. (30 June 1905), 17 (10): 891-921.

26. Minkowski, Hermann (1909). "Raum und Zeit". Jahresbericht der Deutschen Mathematiker-Vereinigung. 18: 75-88

27. Albert Einstein: "Die Feldgleichungen der Gravitation". Sitzungsberichte der Preussischen Akademie der Wissenschaften zu Berlin, S. 844-847, 25. November 1915

28. Albert Einstein. "Kosmologische Betrachtungen zur allgemeinen Relativitätstheorie". (1917) Sitzungsberichte der Königlich Preußischen Akademie der Wissenschaften. part 1. Berlin, DE: 142-152.

29. Alexander Friedmann. "Über die Krümmung des Raumes". (1922) Z. Phys. (in German). 10 (1): 377-386

30. Peebles, P. J. E.; Ratra, Bharat (2003). "The cosmological constant and dark energy". Reviews of Modern Physics. 75 (2). American Physical Society: 559-606

31. Riess, Adam G.; Filippenko; Challis; Clocchiatti; Diercks; Garnavich; Gilliland; Hogan; Jha; Kirshner; Leibundgut; Phillips; Reiss; Schmidt; Schommer; Smith; Spyromilio; Stubbs; Suntzeff; Tonry (1998). "Observational evidence from supernovae for an accelerating universe and a cosmological constant". Astronomical Journal. 116 (3): 1009-1038.

32. Perlmutter, S.; Aldering; Goldhaber; Knop; Nugent; Castro; Deustua; Fabbro; Goobar; Groom; Hook; Kim; Kim; Lee; Nunes; Pain; Pennypacker; Quimby; Lidman; Ellis; Irwin; McMahon; Ruiz-Lapuente; Walton; Schaefer; Boyle; Filippenko; Matheson; Fruchter; et al. (1999). "Measurements of Omega and Lambda from 42 high redshift supernovae". Astrophysical Journal. 517 (2): 565-586.

33. arXiv:astro-ph/0501171

34. Fritz Zwicky, 1933, Helvetica Phys. Acta,6,110

35. V. Rubin and K. Ford, Astrophysical Journal, vol. 159, p.379 (February 1970).

36. A. A. Penzias and R. W. Wilson, "A Measurement of excess antenna temperature at 4080-Mc/s, Astrophys. J." 142 (1965) 419.

37. https://www.jpl.nasa.gov/news/planck-mission-brings-universe-into-sharp-focus

38. D. J. Fixsen, "The Temperature of the Cosmic Microwave Background". (2009) The Astrophysical Journal. 707 (2): 916-920.

39. Georges Lemaître, "A Homogeneous Universe of Constant Mass and Growing Radius Accounting for the Radial Velocity of Extragalactic Nebulae," (1927) Annales Soc. Sci. Bruxelles A 47- 49.

40. Bennett, C. L.; et al. (2011). "Seven-Year Wilkinson Microwave Anisotropy Probe (WMAP) Observations: Are There Cosmic Microwave Background Anomalies?". Astrophysical Journal Supplement Series. 192 (2): 17

41. arXiv:2002.06892 [astro-ph.CO]

42. Starobinsky, A.A. (December 1979). "Spectrum of relict gravitational radiation and the early state of the universe". Journal of Experimental and Theoretical Physics Letters. 30: 682

43. Guth, A. H. 1981, "The Inflationary Universe: A Possible Solution To The Horizon And Flatness Problems, Phys. Rev. D, 23, 347

44. A.D. Linde, Chaotic inflation, Physics Letters B, Volume 129, Issues 3-4, 1983, Pages 177-181

45. J. R. Gott III, et. al., "A Map of the Universe," Astronomical Journal, vol. 624, pp. 463-484, 2005

46. https://www.seti.org/frank-d-drake-1930-2022.

47. The first appearance of the Drake Equation in print, detail of page 324, in "The radio search for intelligent extraterrestrial life," by Frank Drake, in Current Aspects of Exobiology, ed. by G. Mamikunian and M.H. Briggs, 1965 (Linda Hall Library)

48. Wilson, T.L. (1984) Bayes' Theorem and the Real SETI Equation. Quat. J. Royal Astr. Soc. 25: 435-448.

49. Glade N, Ballet P, Bastien O. A stochastic process approach of the drake equation parameters. International Journal of Astrobiology. 2012;11(2):103-108.

50. N. Prantzos. A probabilistic analysis of the Fermi paradox regarding the Drake formula: the role of the L factor. Monthly Notices of the Royal Astronomical Society, 2020, 493 (3), pp.3464- 3472.

51. arXiv:2105.03984 [physics.pop-ph]

52. arXiv:2311.05390 [astro-ph.EP]

53. https://en.wikipedia.org/wiki/Drake_equation

54. Martin N.F.; et al. (2004). "A dwarf galaxy remnant in Canis Major: The fossil of an in-plane accretion on to the Milky Way". Monthly Notices of the Royal Astronomical Society. 348 (1): 12-23.

55. "Dissolving the Fermi Paradox, arXiv:1806.02404 [physics. Pop-ph].

56. https://www.space.com/17137-how-hot-is-the-sun.html

57. Cannon, Annie Jump; Pickering, Edward Charles (1912). "Classification of 1,688 southern stars by means of their spectra". Annals of the Astronomical Observatory of Harvard College. 56 (5): 115.

58. Title: A stars as physics laboratories, Authors: Landstreet, J. D. Journal: The A-Star Puzzle, held in Poprad, Slovakia, July 8-13, 2004. Edited by J. Zverko, J. Ziznovsky, S.J. Adelman, and W.W. Weiss, IAU Symposium, No. 224. Cambridge, UK: Cambridge University Press, 2004., p.423-432.

59. Landstreet JD. A stars as physics laboratories. Proceedings of the International Astronomical Union. 2004;2004(IAUS224):423-432.
60. F. Cruz Aguirre et al 2023 ApJ 956 79.
61. https://www.space.com/habitable-planets-common-sunlike-stars-milky-way
62. arXiv:2012.02061 [astro-ph.SR]
63. Bardeen, J. M., 1973, "Rapidly rotating stars, disks, and black holes", in Black Holes, ed. C. DeWitt and B. S. DeWitt, Gordon and Breach, New York, 241.
64. Shipman, H. L.; Yu, Z; Du, Y.W (1975), "The implausible history of triple star models for Cygnus X-1 Evidence for a black hole", Astrophysical Letters, 16 (1): 9-12.
65. Francisco Nogueras-Lara et al 2021 ApJ 920 97.
66. "Slow Star Formation in the Milky Way: Theory Meets Observations," Neal J. Evans II et al 2022 ApJL 929 L18.
67. Nutman, Allen P.; Bennett, Vickie C.; Friend, Clark R.L.; et al. (September 22, 2016). "Rapid emergence of life shown by discovery of 3,700-million-year-old microbial structures". Nature. 537 (7621): 535-538.
68. Wandel, A. Habitability and sub glacial liquid water on planets of M-dwarf stars. Nat Commun 14, 2125 (2023).
69. Rampelotto PH. Extremophiles and extreme environments. Life (Basel). 2013 Aug 7;3(3):482-5.
70. D'Amico S, Collins T, Marx JC, Feller G, Gerday C. Psychrophilic microorganisms: challenges for life. EMBO Rep. 2006 Apr;7(4):385-9.
71. Das, T. et al. (2022). Halophilic, Acidophilic, Alkaliphilic, Metallophilic, and Radioresistant Fungi: Habitats and Their Living Strategies. In: Sahay, S. (eds) Extremophilic Fungi. Springer, Singapore.
72. https://www.nationalgeographic.com/animals/invertebrates/facts/tardigrades-water-bears
73. Cassan, A., Kubas, D., Beaulieu, JP. et al. One or more bound planets per Milky Way star from microlensing observations. Nature 481, 167-169 (2012).
74. https://www.esa.int/Science_Exploration/Space_Science/Juice
75. Paganini, L., Villanueva, G.L., Roth, L. et al. A measurement of water vapour amid a largely quiescent environment on Europa. Nat Astron 4, 266-272 (2020).
76. Candice J. Hansen et al. ,Enceladus' Water Vapor Plume.Science311,1422-1425(2006).https://doi.org/10.1126/science.1121254
77. V. Cottini, C.A. Nixon, D.E. Jennings, C.M. Anderson, N. Gorius, G.L. Bjoraker, A. Coustenis, N.A. Teanby, R.K. Achterberg, B. Bézard, R. de Kok, E. Lellouch, P.G.J. Irwin, F.M. Flasar, G. Bampasidis, Water vapor in Titan's stratosphere from Cassini CIRS far-infrared spectra, Icarus, Volume 220, Issue 2, 2012, Pages 855-862,
78. Anesio, A.M., Lutz, S., Chrismas, N.A.M. et al. The microbiome of glaciers and ice sheets. Biofilms Microbiomes 3, 10 (2017).
79. Ate H Jaarsma, Katie Sipes, Athanasios Zervas, Francisco Campuzano Jiménez, Lea Ellegaard-Jensen, Mariane S Thøgersen, Peter Stougaard, Liane G Benning, Martyn Tranter, Alexandre M Anesio, Exploring microbial diversity in

Greenland Ice Sheet supraglacial habitats through culturing-dependent and - independent approaches, FEMS Microbiology Ecology, Volume 99, Issue 11, November 2023, fiad119.

80. Oba, Y., Takano, Y., Furukawa, Y. et al. Identifying the wide diversity of extraterrestrial purine and pyrimidine nucleobases in carbonaceous meteorites. Nat Commun 13, 2008 (2022).

81. Hollinger, Maik (2016). "Life from Elsewhere - Early History of the Maverick Theory of Panspermia". Sudhoffs Archiv. 100 (2): 188-205.

82. Kumar D, Steele EJ, Wickramasinghe NC. Preface: The origin of life and astrobiology. Adv Genet. 2020;106:xv-xviii.

83. Siraj, Amir; Loeb, Avi (20 September 2022). "Interstellar Meteors are Outliers in Material Strength". The Astrophysical Journal. 941 (2): L28.

84. McKay CP. Requirements and limits for life in the context of exoplanets. Proc Natl Acad Sci U S A. 2014 Sep 2;111(35):12628-33.

85. Hunten, Donald M.; Shemansky, Donald Eugene; Morgan, Thomas Hunt (1988). "The Mercury atmosphere". In Vilas, Faith; Chapman, Clark R.; Shapley Matthews, Mildred (eds.).

86. Gomez-Leal, Illeana; Kaltenegger, Lisa; Lucarini, Valerio; Lunkeit, Frank (2019). "Climate sensitivity to ozone and its relevance on the habitability of Earth-like planets". Icarus. 321: 608-618.

87. Leconte, Jeremy; Forget, Francois; Charnay, Benjamin; Wordsworth, Robin; Pottier, Alizee (2013). "Increased insolation threshold for runaway greenhouse processes on Earth-like planets". Nature. 504 (7479): 268-71.

88. Ramirez, Ramses; Kaltenegger, Lisa (2017). "A Volcanic Hydrogen Habitable Zone". The Astrophysical Journal Letters. 837 (1).

89. Pierrehumbert, Raymond; Gaidos, Eric (2011). "Hydrogen Greenhouse Planets Beyond the Habitable Zone". The Astrophysical Journal Letters. 734 (1).

90. https://spaceplace.nasa.gov/solar-cycles/en/

91. Bochanski, J. J., Hawley, S. L., Covey, K. R., et al. 2010, AJ, 139, 2679.

92. Aomawa L. Shields, Sarah Ballard, John Asher Johnson, The habitability of planets orbiting M-dwarf stars, Physics Reports, Volume 663, 2016, Pages 1-38.

93. Lorenz, Edward N. (1963). "Deterministic non-periodic flow". Journal of the Atmospheric Sciences. 20 (2): 130-141

94. A computer assisted proof for 100,000 years stability of the solar system" by Angel Zhivkov and Ivaylo Tounchev, arXiv preprint: 2206.13467 (2022)

95. Michelle L. Hill et al 2023 AJ 165 34.

96. https://www.frontiersin.org/journals/astronomy-and-space-sciences/articles/10.3389/fspas.2024.1372057/full

97. J.B. Pollack, J.F. Kasting, S.M. Richardson, K. Poliakoff, "The case for a wet, warm climate on early Mars", Icarus, Volume 71, Issue 2,1987, Pages 203-224.

98. Friedrich Wöhler (1828). "Ueber künstliche Bildung des Harnstoffs". Annalen der Physik und Chemie. 88 (2): 253-256

99. See: Kolbe, H. (1845). "Beiträge zur Kenntniß der gepaarten Verbindungen" [Contributions to [our] knowledge of paired compounds]. Annalen der Chemie und Pharmacie (in German). 54 (2): 145-188.

100. Yamashita H. Biological Function of Acetic Acid-Improvement in Obesity and Glucose Tolerance by Acetic Acid in Type 2 Diabetic Rats. Crit Rev Food Sci Nutr. 2016 Jul 29;56 Suppl 1:S171-5.

101. Miller, Stanley L. (1953). "Production of Amino Acids Under Possible Primitive Earth Conditions". Science. 117 (3046): 528-9.

102. Lopez MJ, Mohiuddin SS. Biochemistry, Essential Amino Acids. [Updated 2024 Apr 30]. In: StatPearls [Internet]. Treasure Island (FL): StatPearls Publishing; 2024 Jan.

103. Daniel S. Helman, Galactic distribution of chirality sources of organic molecules, Acta Astronautica, Volume 151, 2018, Pages 595-602.

104. Pasteur L. Memoires sur la relation qui peut exister entre la forme crystalline et al composition chimique, et sur la cause de la polarization rotatoire. C R Acad Sci. 1848;26:535–538.

105. Cartus AT. 2012. D-Amino acids and cross-linked amino acids as food contaminants. In: Schrenk D, editor. Chemical contaminants and residues in food. Cambridge: Woodhead Publishing; p. 286-319.

106. Das K, Balaram H, Sanyal K. Amino Acid Chirality: Stereospecific Conversion and Physiological Implications. ACS Omega. 2024 Jan 26;9(5):5084-5099.

107. Fujii N. D-amino acids in living higher organisms. Orig Life Evol Biosph. 2002 Apr;32(2):103-27.

108. Chemical Complexity and Consequent Homochirality from Weak Neutral Currents Acting upon Paramagnetic Enantiomers. Stevenson, Cheryl D., Davis, John P., (2021). American Chemical Society.

109. Experimental Test of Parity Conservation in Beta Decay C. S. Wu, E. Ambler, R. W. Hayward, D. D. Hoppes, and R. P. Hudson Phys. Rev. 105, 1413 - Published 15 February 1957.

110. Weller, Michael G. 2024. "The Mystery of Homochirality on Earth" Life 14, no. 3: 341.

111. S. Furkan Ozturk et al. , Origin of biological homochirality by crystallization of an RNA precursor on a magnetic surface.Sci. Adv.9,eadg8274(2023).

112. Koga, T., Naraoka, H. A new family of extraterrestrial amino acids in the Murchison meteorite. Sci Rep 7, 636 (2017).

113. Bada, Jeffrey L.; Cronin, John R.; Ho, Ming-Shan; Kvenvolden, Keith A.; Lawless, James G.; Miller, Stanley L.; Oro, J.; Steinberg, Spencer (10 February 1983). "On the reported optical activity of amino acids in the Murchison meteorite". Nature. 301 (5900): 494-496.

114. Gomes, R., Levison, H., Tsiganis, K. et al. Origin of the cataclysmic Late Heavy Bombardment period of the terrestrial planets. Nature 435, 466-469 (2005).

115. Giovanni Picardi et al. , Radar Soundings of the Subsurface of Mars.Science310,1925-1928(2005).

116. Cooper, G.M. (2000) The Cell A Molecular Approach. 2nd Edition, Sunderland (MA) Sinauer Associates, The Development and Causes of Cancer.

117. Moody, E.R.R., Álvarez-Carretero, S., Mahendrarajah, T.A. et al. The nature of the last universal common ancestor and its impact on the early Earth system. Nat Ecol Evol (2024).

118. Knoll AH. Paleobiological perspectives on early eukaryotic evolution. Cold Spring Harb Perspect Biol. 2014 Jan 1;6(1):a016121.

119. https://www.rhs.org.uk/science/articles/fun-fungi-facts

120. Maloof, A. C.; Porter, S. M.; Moore, J. L.; Dudas, F. O.; Bowring, S. A.; Higgins, J. A.; Fike, D. A.; Eddy, M. P. (2010). "The earliest Cambrian record of animals and ocean geochemical change". Geological Society of America Bulletin. 122 (11-12): 1731-1774.

121. Butterfield, N. J. (1 February 2003). "Exceptional Fossil Preservation and the Cambrian Explosion". Integrative and Comparative Biology. 43 (1): 166-177.

122. Conway Morris, Simon (1979). "The Burgess Shale (Middle Cambrian) Fauna". Annual Review of Ecology and Systematics. 10: 327-349.

123. Walter G. Kühne, "On a Triconodont tooth of a new pattern from a Fissure-filling in South Glamorgan", Proceedings of the Zoological Society of London, volume 119 (1949-1950) pages 345-350

124. Robert D. Martin, Primates, Current Biology, Volume 22, Issue 18, 2012, Pages R785-R790,

125. Stevens NJ, Seiffert ER, O'Connor PM, Roberts EM, Schmitz MD, Krause C, Gorscak E, Ngasala S, Hieronymus TL, Temu J. Palaeontological evidence for an Oligocene divergence between Old World monkeys and apes. Nature. 2013 May 30;497(7451):611-4.

126. https://australian.museum/learn/science/human-evolution/humans-are-apes-great-apes/

127. Patterson N, Richter DJ, Gnerre S, Lander ES, Reich D (2006). "Genetic evidence for complex speciation of humans and chimpanzees". Nature. 441 (7097): 1103-08.

128. Carotenuto F, Tsikaridze N, Rook L, Lordkipanidze D, Longo L, Condemi S, Raia P. Venturing out safely: The biogeography of Homo erectus dispersal out of Africa. J Hum Evol. 2016 Jun;95:1-12.

129. Raup DM, Sepkoski JJ (March 1982). "Mass extinctions in the marine fossil record". Science. 215 (4539): 1501-1503.

130. Jing, X., Yang, Z., Mitchell, R.N., et al. Ordovician-Silurian true polar wander as a mechanism for severe glaciation and mass extinction. Nat Commun 13, 7941 (2022).

131. Robert W. Gess, Per E. Ahlberg, A high latitude Gondwanan species of the Late Devonian tristichopterid Hyneria (Osteichthyes: Sarcopterygii), PLOS ONE, 18, 2, (e0281333), (2023).

132. Hautmann, Michael. (2012). Extinction: End-Triassic Mass Extinction. Encyclopedia of Life Sciences.

133. Thomas Servais, David A. T. Harper, Björn Kröger, Christopher Scotese, Alycia L. Stigall, Yong-Yi Zhen, Changing palaeobiogeography during the Ordovician Period, Geological Society, London, Special Publications, 10.1144/SP532-2022-168, 532, 1, (2022).

134. Qiu, Zhen; Zou, Caineng; Mills, Benjamin J. W.; Xiong, Yijun; Tao, Huifei; Lu, Bin; Liu, Hanlin; Xiao, Wenjiao; Poulton, Simon W. (5 April 2022). "A nutrient control on expanded anoxia and global cooling during the Late Ordovician mass extinction". Communications Earth and Environment. 3 (1): 82.

135. https://www.pbs.org/wgbh/evolution/change/deeptime/ordovician.html

136. Harper DAT. Late Ordovician Mass Extinction: Earth, fire and ice. Natl Sci Rev. 2023 Dec 18;11(1).

137. arXiv:astro-ph/0309415

138. Ball, P. Gamma-ray burst linked to mass extinction. Nature (2003).

139. Frederiks, D.; Svinkin, D.; Lysenko, A. L.; Molkov, S.; Tsvetkova, A.; Ulanov, M.; Ridnaia, A.; Lutovinov, A. A.; Lapshov, I.; Tkachenko, A.; Levin, V. (May 18, 2023). "Properties of the Extremely Energetic GRB 221009A from Konus-WIND and SRG/ART-XC Observations". The Astrophysical Journal Letters. 949 (1): L7.

140. Becker, R. T.; Marshall, J. E. A.; Da Silva, A. -C.; Agterberg, F. P.; Gradstein, F. M.; Ogg, J. G. (1 January 2020), Gradstein, Felix M.; Ogg, James G.; Schmitz, Mark D.; Ogg, Gabi M. (eds.), "Chapter 22 - The Devonian Period", Geologic Time Scale 2020, Elsevier, pp. 733-810.

141. Cocks, L. Robin M.; Torsvik, Trond H., eds. (2016), "Devonian", Earth History and Palaeogeography, Cambridge: Cambridge University Press.

142. https://new.nsf.gov/news/climate-change-factors-fossil-record-accelerate

143. George R. McGhee, Chapter 3Modelling late Devonian extinction hypotheses, Editor(s): D.J. Over, J.R. Morrow, P.B. Wignall, Developments in Palaeontology and Stratigraphy, Elsevier, Volume 20, 2005.

144. Fields BD, Melott AL, Ellis J, Ertel AF, Fry BJ, Lieberman BS, Liu Z, Miller JA, Thomas BC. Supernova triggers for end-Devonian extinctions. Proc Natl Acad Sci USA. 2020 Sep 1;117(35):21008-21010.

145. Erwin, D. H. The Permo-Triassic extinction. Nature 367, 231-236 (1994).

146. Burgess, S. D., Bowring, S. and Shen, S. Z. High-precision timeline for Earth's most severe extinction. Proc. Natl Acad. Sci. USA 111, 3316-3321 (2014).

147. Wang J, Pfefferkorn HW. Nystroemiaceae, a new family of Permian gymnosperms from China with an unusual combination of features. Proc Biol Sci. 2010 Jan 22;277(1679):301-9.

148. https://www.livescience.com/43219-permian-period-climate-animals-plants.html

149. arXiv:1911.10188 [physics.ao-ph]

150. K Kaiho, Y Kajiwara, T Nakanao, Y Miura, H Kawahata, K Tazaki, M Ueshima, Z Chen, G R Shi Geology 29, 815-818 (2001).

151. Whiteside, Jessica H.; Olsen, Paul E.; Eglington, Timothy; Brookfield, Michael E.; Sambrotto, Raymond N. (22 March 2010). "Compound-specific carbon isotopes from Earth's largest flood basalt eruptions directly linked to the end-Triassic mass extinction". Proceedings of the National Academy of Sciences of the United States of America. 107 (15): 6721-6725.

152. G.M. Stampfli, C. Hochard, C. Vérard, C. Wilhem, J. vonRaumer, The formation of Pangea, Tectonophysics, Volume 593, 2013, Pages 1-19.

153. L.H. Tanner, S.G. Lucas, M.G. Chapman, Assessing the record and causes of Late Triassic extinctions, Earth-Science Reviews, Volume 65, Issues 1-2, 2004, Pages 103-139,

154. Jablonski, D.; Chaloner, W. G. (1994). "Extinctions in the fossil record (and discussion)". Philosophical Transactions of the Royal Society of London B. 344 (1307): 11-17.

155. Alvarez, Luis W.; Alvarez, Walter; Asaro, F.; Michel, H. V. (1980). "Extraterrestrial cause for the Cretaceous-Tertiary extinction" (PDF). Science. 208 (4448): 1095-1108.

156. Donovan, M., Iglesias, A., Wilf, P. et al. Rapid recovery of Patagonian plant-insect associations after the end-Cretaceous extinction. Nat Ecol Evol 1, 0012 (2017).

157. https://samnoblemuseum.ou.edu/understanding-extinction/minor-mass-extinctions/

158. The Zoologist Guide to the Galaxy. What Animals on Earth Reveal About Aliens - and Ourselves. Dr. Arik Kershenbaum. Penguin Books. (2020).

159. Kardashev, N.S. (1964) Transmission of Information by Extraterrestrial Civilizations. Soviet Astronomy, 8, 217-221.

160. Zhang, A., Yang, J., Luo, Y. et al. Forecasting the progression of human civilization on the Kardashev Scale through 2060 with a machine-learning approach. Sci Rep 13, 11305 (2023).

161. https://mkaku.org/home/articles/the-physics-of-extraterrestrial-civilizations/

162. Seager S, Bains W, Petkowski JJ. Toward a List of Molecules as Potential Biosignature Gases for the Search for Life on Exoplanets and Applications to Terrestrial Biochemistry. Astrobiology. 2016 Jun;16(6):465-85.

163. arXiv:1801.00732 [astro-ph.SR]

164. Woods, P. More dips for Tabby's Star. Nat Astron 2, 185 (2018).

165. J. Richard Gott, III (1993). "Implications of the Copernican principle for our future prospects." Nature. 363 (6427): 315-319.

166. Callaway, Ewan (7 June 2017). "Oldest Homo sapiens fossil claim rewrites our species' history." Nature.

167. Chris Stringer, Julia Galway-Witham. When did modern humans leave Africa? Science 359, 389-390 (2018).

168. Romanovskaya IK. Migrating extraterrestrial civilizations and interstellar colonization: implications for SETI and SETA. International Journal of Astrobiology. 2022;21(3):163-187.

169. arXiv:2110.15213 [physics.pop-ph]

170. https://www.seti.org

171. https://www.seti.org/press-release/massive-radio-array-search-extraterrestrial-signals-other-civilizations

172. George Gaylord Simpson. The Nonprevalence of Humanoids. Science, 21 Feb 1964, Vol 143, Issue 3608, pp. 769-775

173. Ernst Mayr "A Critique of the Search for Extraterrestrial Intelligence", The Bioastronomy News, vol. 7, no. 3, 1995.

174. Rare Earth - why complex life is uncommon in the universe. Peter Ward and Donald Brownlee Copernicus Publishers, New York, USA (2000)

175. Lucky planet: why Earth is exceptional– and what that means for life in the universe. David Waltham. New York : Basic Books, a member of the Perseus Books Group, [2014].

176. Laskar, J, Chaotic Diffusion in the Solar System, https://arxiv.org/pdf/0802.3371

177. Aleksandr Lyapunov, The General Problem of Stability of Motion, 1892, Kharkov University

178. Tokovinin, A. 2014a, AJ, 147, 86

179. M. Cuntz 2014 ApJ 780 14

180. Elisa V. Quintana, Jack J. Lissauer arXiv:0705.3444 [astro-ph]

181. Wiegert, Paul A.; Holman, Matt J. (April 1997). "The stability of planets in the Alpha Centauri system". The Astronomical Journal. 113 (4): 1445-1450.

182. Jerome A. Orosz et al. 2019 AJ 157 174

183. Reipurth, B. and Mikkola, S. Formation of the widest binary stars from dynamical unfolding of triple systems. Nature 492, 221-224 (2012).

184. Mamajek, E.E.; Hillenbrand, L.A. (2008). "Improved age estimation for Solar-type dwarfs using activity-rotation diagnostics". Astrophysical Journal. 687 (2): 1264-1293.

185. Schwamb, Megan E.; Orosz, Jerome A.; Carter, Joshua A.; Welsh, William F.; Fischer, Debra A.; Torres, Guillermo; Howard, Andrew W.; Crepp, Justin R.; Keel, William C.; Lintott, Chris J.; Kaib, Nathan A.; Terrell, Dirk; Gagliano, Robert; Jek, Kian J.; Parrish, Michael; Smith, Arfon M.; Lynn, Stuart; Simpson, Robert J.; Giguere, Matthew J.; Schawinski, Kevin (2013). "Planet Hunters: A Transiting Circumbinary Planet in a Quadruple Star System". The Astrophysical Journal. 768 (2): 127.

186. https://www.nasa.gov/wp-content/uploads/2015/01/yoss_act_4.pdf

187. https://www.nasa.gov/wp-content/uploads/2015/01/yoss_act_4.pdf

188. https://science.nasa.gov/exoplanets/trappist1/

189. de Wit, Julien; Wakeford, Hannah R.; et al. (5 February 2018). "Atmospheric reconnaissance of the habitable-zone Earth-sized planets orbiting TRAPPIST-1". Nature Astronomy. 2 (3). Nature: 214-219.

190. Måns Wallner et al. , Abiotic molecular oxygen production-Ionic pathway from sulfur dioxide.Sci. Adv.8,eabq5411(2022).

191. Liu, Zibo, and Dongdong Ni. "Quantitative correlation of refractory elemental abundances between rocky exoplanets and their host stars." Astronomy and Astrophysics 674 (2023): A137.

192. Alan P. Boss Metallicity and Planet Formation: Models, August 2009Proceedings of the International Astronomical Union 5(S265):391 - 398.

193. Pineda, J.S., Villadsen, J. Coherent radio bursts from known M-dwarf planet-host YZ Ceti. Nat Astron 7, 569-578 (2023).

194. https://www.nsf.gov/mps/ast/outreach/MilkyWayLEDCraft2019.pdf

195. arXiv:1810.12641.

196. GALACTICNUCLEUS: A high-angular-resolution JHKs imaging survey of the Galactic centre - II. First data release of the catalogue and the most detailed CMDs of the GC. F. Nogueras-Lara, R. Schödel, A. T. Gallego-Calvente, H. Dong, E. Gallego-Cano, B. Shahzamanian, J. H. V. Girard, S. Nishiyama, F. Najarro and N. Neumayer. AandA, 631 (2019) A20.

197. Daly Reginald A. 1946 Origin of the Moon and its topography. Proc. Am. Phil. Soc. 90, 104-119.

198. The deep Earth may not be cooling down", by Denis Andrault, Julien Monteux, Michael Le Bars and Henri Samuel Earth and Planetary Science Letters, 30 mars 2016.

199. Nakajima, M., Genda, H., Asphaug, E. et al. Large planets may not form fractionally large moons. Nat Commun 13, 568 (2022).

200. https://science.nasa.gov/solar-system/oort-cloud/

201. ...Janson M, Gratton R, Rodet L, Vigan A, Bonnefoy M, Delorme P, Mamajek EE, Reffert S, Stock L, Marleau GD, Langlois M, Chauvin G, Desidera S, Ringqvist S, Mayer L, Viswanath G, Squicciarini V, Meyer MR, Samland M, Petrus S, Helled R, Kenworthy MA, Quanz SP, Biller B, Henning T, Mesa D, Engler N, Carson JC. A wide-orbit giant planet in the high-mass b Centauri binary system. Nature. 2021 Dec;600(7888):231-234. https://doi.org/10.1038/s41586-021-04124-8.

202. Alfred Wegener, "The Origin of Continents". (1912).

203. https://education.nationalgeographic.org/resource/plate-tectonics/

204. https://www.whoi.edu/multimedia/carbon-dioxide-shell-building-and-ocean-acidification/

205. Weller, M.B., Evans, A.J., Ibarra, D.E. et al. Venus's atmospheric nitrogen explained by ancient plate tectonics. Nat Astron 7, 1436-1444 (2023).

206. Kattenhorn, Simon and Prockter, Louise. (2014). Evidence for subduction in the ice shell of Europa. Nature Geoscience. 7. 762-767.

207. Tobias G. Meier et al 2021 ApJL 908 L48.

208. https://www.nationalgeographic.com/animals/invertebrates/facts/tardigrades-water-bears

209. Hanson, Robin, The Great Filter - Are We Almost Past It?

210. Schrödinger, Erwin (1974). What is life? mind and matter: the physical aspect of the living cell. Cambridge University Press.

211. Watson, J., Crick, F. Molecular Structure of Nucleic Acids: A Structure for Deoxyribose Nucleic Acid. Nature 171, 737-738 (1953).

212. Libretti S, Puckett Y. Physiology, Homeostasis. [Updated 2023 May 1]. In: StatPearls [Internet]. Treasure Island (FL): StatPearls Publishing; 2024 Jan.

213. Wang D, Farhana A. Biochemistry, RNA Structure. [Updated 2023 Jul 29]. In: StatPearls [Internet]. Treasure Island (FL): StatPearls Publishing; 2024 Jan-.

214. Pavlinova P, Lambert CN, Malaterre C, Nghe P. Abiogenesis through gradual evolution of autocatalysis into template-based replication. FEBS Lett. 2023 Feb;597(3):344-379.

215. Schrum JP, Zhu TF, Szostak JW. The origins of cellular life. Cold Spring Harb Perspect Biol. 2010 Sep;2(9):a002212.

216. Gánti, Tibor (31 December 2003). Chemoton Theory: Theory of Living Systems. Translated by Elisabeth Csárán. Kluwer Academic/Plenum Publishers.

217. https://www.nature.com/scitable/definition/prokaryote-procariote-18

218. Martin D. Brasier, Owen R. Green, Andrew P. Jephcoat, Annette K. Kleppe, Martin J. Van Kranendonk, John F. Lindsay, Andrew Steele and Nathalie V. Grassineauk. Nature I Vol 416 I 7 March 2002.

219. Bar-On YM, Phillips R, Milo R. The biomass distribution on Earth. Proc Natl Acad Sci U S A. 2018 Jun 19;115(25):6506-6511.

220. Dey G, Thattai M, Baum B. On the Archaeal Origins of Eukaryotes and the Challenges of Inferring Phenotype from Genotype. Trends Cell Biol. 2016 Jul;26(7):476-485.

221. Cavalier-Smith T. Origin of mitochondria by intracellular enslavement of a photosynthetic purple bacterium. Proc Biol Sci. 2006 Aug 7;273(1596):1943-52.

222. Cooper GM. The Cell: A Molecular Approach. 2nd edition. Sunderland (MA): Sinauer Associates; 2000. The Origin and Evolution of Cells.

223. Goodenough U, Heitman J. Origins of eukaryotic sexual reproduction. Cold Spring Harb Perspect Biol. 2014 Mar 1;6(3):a016154.

224. J.T Bonner The Evolution of Complexity by Means of Natural Selection Princeton University Press, Princeton, NJ (1988)

225. Sanderson, M. J. Molecular data from 27 proteins do not support a Precambrian origin of land plants. American Journal of Botany 90, 954-956 (2003).

226. https://www.science.org/content/article/earth-s-first-animals-may-have-been-sea-sponges

227. Herries AI, Martin JM, Leece AB, Adams JW, Boschian G, Joannes-Boyau R, et al. (April 2020). "Contemporaneity of Australopithecus, Paranthropus, and early Homo erectus in South Africa". Science. 368 (6486).

228. https://gorillafund.org/dian-fossey/social-groups/

229. Johnson RT, O'Neill MC, Umberger BR. The effects of posture on the three-dimensional gait mechanics of human walking compared to walking in bipedal chimpanzees. J Exp Biol. 2022 Mar 1;225(5):jeb243272.

230. Gould SJ. 1990. Wonderful life: the burgess shale and the nature of history. New York, NY: W.W. Norton and Co.

231. Bostrom, Nick (2018). The Vulnerable World Hypothesis.

232. Nick Bostrom. Where are They? Why I hope that the search for extraterrestrial life finds nothing. MIT Technology Review, May/June issue (2008): pp. 72-77.

233. Tarter, J. C., Agrawal, A., Ackermann, R., et al. 2010, in Proc. SPIE, Vol. 7819, Instruments, Methods, and Missions for Astrobiology XIII, 781902

234. Hemming, John (1987). The Conquest of the Incas. Penguin Books.

235. John A. Ball, The Zoo Hypothesis, Icarus, Volume 19, Issue 3, 1973, Pages 347-349.

236. Duncan Forgan, arXiv:1105.2497 [astro-ph.EP].

237. George K. Zipf (1935): The Psychobiology of Language. Houghton-Mifflin.

238. Brin, G. D. (1 September 1983). "The Great Silence - the Controversy Concerning Extraterrestrial Intelligent Life". Quarterly Journal of the Royal Astronomical Society. 24: 283-309.

239. https://www.projectnash.com/aliens-the-fermi-paradox-and-the-dark-forest-theory/

240. https://www.ligo.org/science/Publication-GW150914/

241. arXiv:2402.15707v1 [eess.SP] 24 Feb 2024, A Quick Guide to Quantum Communication. Rohit Singh, Member, IEEE, Roshan M. Bodile, Member, IEEE.

242. Einstein, A; B Podolsky; N Rosen (1935-05-15). "Can Quantum-Mechanical Description of Physical Reality be Considered Complete?" (PDF). Physical Review. 47 (10): 777-780.

243. Albert Einstein, Lens-Like Action of a Star by the Deviation of Light in the Gravitational Field. Science84,506-507(1936).

244. https://www.wildlifehotline.com/help/opossums

245. Bates H.W. 1863. The Naturalist on the River Amazons. 2 vols, Murray, London.

246. https://nationalzoo.si.edu/animals/sinaloan-milksnake

247. Emily A. Gilbert, Andrew Vanderburg, Joseph E. Rodriguez, Benjamin J. Hord, Matthew S. Clement, Thomas Barclay, Elisa V. Quintana, Joshua E. Schlieder, Stephen R. Kane, Jon M. Jenkins, Joseph D. Twicken, Michelle Kunimoto, Roland Vanderspek, Giada N. Arney, David Charbonneau, Maximilian N. Günther, Chelsea X. Huang, Giovanni Isopi, Veselin B. Kostov, Martti H. Kristiansen, David W. Latham, Franco Mallia, Eric E. Mamajek, Ismael Mireles, Samuel N. Quinn, George R. Ricker, Jack Schulte, S. Seager, Gabrielle Suissa, Joshua N. Winn, Allison Youngblood, Aldo Zapparata, A Second Earth-sized Planet in the Habitable Zone of the M Dwarf, TOI-700, The Astrophysical Journal Letters, 944, 2, (L35), (2023).

248. That is not dead which can eternal lie: the aestivation hypothesis for resolving Fermi's paradox: arXiv:1705.03394 [physics. pop-ph]

249. Bennett, C.H., Hanson, R. and Riedel, C.J. Comment on "The Aestivation Hypothesis for Resolving Fermi's Paradox'. Found Phys 49, 820-829 (2019)

250. Wolfe JW, Rummel JD. Long-term effects of microgravity and possible countermeasures. Adv Space Res. 1992;12(1):281-4.

251. https://nssdc.gsfc.nasa.gov/planetary/factsheet/saturnfact.html

252. https://www.space.com/nasa-parker-solar-probe-fastest-man-made-object-breaks-record

253. https://www.space.com/how-long-does-it-take-to-get-to-the-moon

254. https://www.sciencealert.com/worlds-largest-nuclear-fusion-rocket-engine-begins-construction

255. arXiv:gr-qc/0009013.

256. Albert Einstein and Nathan Rosen, The Particle Problem in the General Theory of Relativity (abstract), Physical Review, vol. 48, n. 1, American Physical Society,1935.

257. https://www.nasa.gov/headquarters/library/find/bibliographies/long-term-challenges-to-human-space-exploration/

258. Saberhagen, Fred (January 1963). "Fortress Ship". Worlds of IF. pp. 96-105.
259. Computing Machinery and Intelligence Author(s): A. M. Turing Source: Mind, New Series, Vol. 59, No. 236 (Oct. 1950), pp. 433-460. Published by Oxford University Press on behalf of the Mind Association.
260. Bickle, John. 2006. "Reducing mind to molecular pathways: explicating the reductionism implicit in current cellular and molecular neuroscience." Synthese, 151, 411-434.
261. Chalmers, D. J. 1995. Facing up to the problem of consciousness. Journal of Consciousness Studies 2: 200-19.
262. Johnson, Steven (2001). Emergence: The Connected Lives of Ants, Brains, Cities, and Software. New York, NY: Scribner.
263. https://www.scientificamerican.com/article/google-engineer-claims-ai-chatbot-is-sentient-why-that-matters/
264. Le avventure di Pinocchio. Storia di un burattino (1883).
265. Theory of self-reproducing automata by Von Neumann, John, Burks, Arthur W. (Arthur Walter), Publication date 1966, Publisher Urbana, University of Illinois Press
266. Kahn J (2006). "Nanotechnology". National Geographic. 2006 (June): 98-119.
267. Binnig G, Rohrer H (1986). "Scanning tunneling microscopy". IBM Journal of Research and Development. 30 (4): 355-369.
268. Von Neumann, J., Morgenstern, O. (1944). Theory of games and economic behavior. Princeton University Press.
269. Nash, John (1950) "Equilibrium points in n-person games" Proceedings of the National Academy of Sciences 36(1):48-49
270. H. A. Bethe, "Ultimate Catastrophe?," Bull. Atomic Scientists 32, No. 6, 36 (1976).
271. B. Koch, M. Bleicher, and H. Stoecker, "Exclusion of Black Hole Disaster Scenarios at the LHC," Phys. Lett. B 672, 71 (2009).
272. Diogenis Laertii Vitae philosophorum edidit Miroslav Marcovich, Stuttgart-Lipsia, Teubner, 1999-2002. Bibliotheca scriptorum Graecorum et Romanorum Teubneriana, vol. 1: Books I-X
273. "The Theory of the Universal Wave Function," (1955)
274. Messiah, Albert: Quantum Mechanics,1965
275. Planck, Max (1901), "Ueber das Gesetz der Energieverteilung im Normalspectrum", Ann. Phys., 309 (3): 553-63
276. Faye, Jan (2019). "Copenhagen Interpretation of Quantum Mechanics". In Zalta, Edward N. (ed.). Stanford Encyclopedia of Philosophy. Metaphysics Research Lab, Stanford University.
277. DeWitt, B. S. (1967). "Quantum Theory of Gravity. I. The Canonical Theory". Phys. Rev. 160 (5): 1113-1148.
278. Pieter Thyssen. The Block Universe: A Philosophical Investigation in Four Dimensions. Philosophy.Institute of Philosophy, KU Leuven (Belgium), 2020
279. Totani, T. Emergence of life in an inflationary universe. Sci Rep 10, 1671 (2020)
280. Guth, Alan H. (2007). "Eternal inflation and its implications". J. Phys. A. 40 (25): 6811-6826.

281. P.J. Steinhardt "Natural Inflation," in The Very Early Universe, ed. by G. Gibbons, S. Hawking and S. Siklos, (Cambridge University Press: 1983), pp. 251

282. "Evolution of the Solar System and the Expansion of the Universe" R. H. Dicke and P. J. E. Peebles, Phys. Rev. Lett. 12, 435 - 1964

283. Just Six Numbers: The Deep Forces That Shape the Universe (Science Masters): Written by Sir Martin Rees, 1999 Edition, (First Edition) Publisher: W and N

284. David J. Chalmers The matrix as metaphysics. 2005 - In Christopher Grau (ed.), Philosophers Explore the Matrix. Oxford University Press. pp. 132

285. Nick Bostrom Are we living in a computer simulation? - 2003 - Philosophical Quarterly 53 (211):243-255.

286. https://iep.utm.edu/brain-in-a-vat-argument

287. Kaluza, T. (1921) Zum Unitätsproblem in der Physik. Sitzungsber. Preuss. Akad. Wiss. Berlin. (Math. Phys.), 966-972.

288. Klein, Oskar (1926). "Quantentheorie und fünfdimensionale Relativitätstheorie". Zeitschrift für Physik A (in German). 37 (12): 895-906.

289. Becker, Katrin; Becker, Melanie; Schwarz, John (2007). String theory and M-theory: A modern introduction. Cambridge University Press.

290. Bergshoeff, Eric; Sezgin, Ergin; Townsend, Paul (1987). "Supermembranes and eleven-dimensional supergravity" (PDF). Physics Letters B. 189 (1): 75-78

291. Westman, Robert S.. "Nicolaus Copernicus". Encyclopedia Britannica, 7 Oct. 2024, https://www.britannica.com/biography/Nicolaus-Copernicus Accessed 14 October 2024.

292. Mandelbrot Benoit (1977) Fractals: Form, chance and dimension. W.H. Freeman, San Francisco

293. P.J.E. Peebles, The fractal galaxy distribution, Physica D: Nonlinear Phenomena, Volume 38, Issues 1-3,1989, Pages 273-278.

294. Dicke, R. Dirac's Cosmology and Mach's Principle. Nature 192, 440-441 (1961)

295. Carter, Brandon. 1974. Large number coincidences and the anthropic principle in cosmology. In Longair (1974), p.291-298.

296. Tintori SC, Çağlar D, Ortiz P, Chyzhevskyi I, Mousseau TA, Rockman MV. Environmental radiation exposure at Chornobyl has not systematically affected the genomes or chemical mutagen tolerance phenotypes of local worms. Proc Natl Acad Sci USA. 2024 Mar 12;121(11):e2314793121.

297. Postberg, F., Sekine, Y., Klenner, F. et al. Detection of phosphates originating from Enceladus's ocean. Nature 618, 489-493 (2023).

298. "Frequently Asked Questions about Europa". NASA. 2012.

299. https://science.nasa.gov/jupiter/moons/europa/europa-facts/

300. Phillips, Cynthia B.; Pappalardo, Robert T. (20 May 2014). "Europa Clipper Mission Concept". Eos, Transactions American Geophysical Union. 95 (20): 165-167.

301. https://europa.nasa.gov/resources/174/induced-magnetic-field-from-europas-subsurface-ocean/

302. Berna, Francesco (2012). "Microstratigraphic evidence of in situ fire in the Acheulean strata of Wonderwerk Cave, Northern Cape province, South Africa". Proceedings of the National Academy of Sciences of the United States of America. 109 (20): E1215-20

303. Locey KJ, Lennon JT. Scaling laws predict global microbial diversity. Proc Natl Acad Sci. 2016;113:5970-5975.

Index